W. Ehrfeld, V. Hessel, H. Löwe

Microreactors

Wolfgang Ehrfeld, Volker Hessel, Holger Löwe

Microreactors

New Technology for Modern Chemistry

WILEY-VCH

Weinheim · New York · Chichester · Brisbane · Singapore · Toronto

W. Ehrfeld, V. Hessel, H. Löwe
Institut für Mikrotechnik Mainz GmbH
Carl-Zeiss-Straße 18–20
55129 Mainz
Germany

First Edition 2000

Cover Illustration: Interdigital micromixers realized by small series fabrication.
Courtesy of IMM Mainz GmbH, Mainz, Germany.

Library of Congress Card No.: applied for

British Library Cataloguing-in-Publication Data:
A catalogue record for this book is available from the British Library

Die Deutsche Bibliothek – CIP-Einheitsaufnahme
A catalogue record for this book is available from Die Deutsche Bibliothek

Composition: Text- und Software-Service Manuela Treindl, D-93059 Regensburg
Printing: Strauss Offsetdruck GmbH, D-69509 Mörlenbach
Bookbinding: Wilh. Osswald & Co., D-67433 Neustadt
Printed in the Federal Republic of Germany

Preface

Today's microreaction technology is no longer in its infancy. Whereas the first four years of development were characterized by basic feasibility work and the development of methods, sometimes even accompanied by skepticism and lack of understanding, the success of this emerging technology these days is beyond doubt. Small microfabricated reactors have proven to provide excellent mass and heat transfer properties as well as uniform flow patterns and residence time distributions. Some of these devices even are available commercially and being tested by a growing community of researchers. This, for instance, holds for interdigital micromixers, thus stimulating a rapid accumulation of know-how in the field of micromixing.

The numerous teams of researchers investigating microreaction technology belong to many disciplines, e.g. chemists, physicists, chemical engineers, material scientists, and mechanical engineers. Meanwhile, a vivid communication has started, the series of "International Conferences on Microreaction Technology" being the focus of this exchange of scientific results. It seems to be that something which was hidden for a long time is now bursting out, and has the potential to a complete change of today's chemical methodologies and, maybe even more notable, the corresponding habits of the researchers.

Industrial contributions played an important role throughout the recent years' developments. Innovative companies like DuPont and BASF promoted and supported these developments by own activities from the very beginning. The Merck company presented their first example of implementing a new industrial process, at least assisted by means of microreaction technology. Axiva seems to be interested to become a professional provider of modern process development utilizing microreactors. A number of companies followed these examples, e.g. Schering, Degussa-Hüls, and Bayer, just to name a few. But not only German companies are involved, Rhône-Poulenc/Rhodia in France, Shell in the Netherlands and the United Kingdom, as well as DuPont and UOP in the U.S.A. have become more and more active.

The technical and scientific development concerning micoreactors is tremendous. Knowledge concerning microfabrication, modeling, design concepts, and testing is provided and increasingly spread between different journals. Due to this growing number of publications and a so far missing common platform, e.g. a journal on microreaction technology, the authors strongly felt that it is high time to summarize and categorize the research work in recent years. In this context, this book presents the state of the art of this new discipline, but is also designated to reveal its full beauty and breathless excitement.

This book is written both for the newcomer and the expert as well as for researchers from industry and research institutions. It is neither intended to attract chemists or chemical engineers only nor being dedicated to "microtechnicians". Hence, the book, as microreaction technology itself, should be of interdisciplinary character in the true sense. It tries to join the different disciplines, to evidence the success of mutual interaction, and to high-light the benefits of such a strategy.

The book contents are subdivided into two major parts. The beginning of each chapter is aimed to present general aspects of a specific class of microdevices, while in separate sections details are discussed therein. Consequently, large parts of this book contain a multitude of single information compiled in a comprehensive volume. Nevertheless, concepts are introduced as well and the respective fields of applications are indicated. Presently, such a systematic analysis in most cases is limited since a number of activities regarding microreactors still remain in their starting phase. However, there is an ultimate need to expand this systematic analysis in a future version of this book.

A book, as all hard and genuine work, seldom is elaborated by a single person or a small group on their own, but rather in a framework of human co-operation. In this context, the authors would like to acknowlegde the aid of all members of the Microreaction Technology and Chemistry Departments at the Institut für Mikrotechnik Mainz. In particular, this refers to L. Agueda for organization, J. Schiewe and Th. Richter for discussion and proof-reading, and Ch. Hofmann for illustration. Additionally, all funders and believers in this technology are acknowlegded, in particular from chemical industries as well as the DECHEMA and DARPA organizations. The same holds for the VCH-Wiley publishers, early recognizing the importance of microreaction technology and providing the possibility for this comprehensive volume. Finally, the authors are deeply indebted to the scientific community active in microreaction technology. For instance, a number of researchers helped a lot by providing illustrations and proof-reading of sections.

IMRET 4 in Atlanta is just around the corner. The technology development certainly will speed up. Standardization and system assembly as well as commercialization of microreactors may become relevant key topics. Production issues based on microreactors – a topic only mentioned in a whisper two years ago – attracts increasing interest. These interesting topics certainly will contribute to the growing progress in microreaction technology.

We are really glad that we actively could take part in such a fascinating development.

W. Ehrfeld, V. Hessel, H. Löwe   Mainz, March 2000

Contents

1	**State of the Art of Microreaction Technology**	1
1.1	Definition	1
1.1.1	Microsystems Termed Microreactor	1
1.1.2	Structural Hierarchy of Microreactors	1
1.1.3	Functional Classification of Microreactors	4
1.1.4	Dividing Line Between Analysis and Reaction Systems	4
1.2	Fundamental Advantages of Microreactors	5
1.2.1	Fundamental Advantages of Miniaturized Analysis Systems	5
1.2.2	Fundamental Advantages of Nano-Scale Reactors	5
1.2.3	Advantages of Microreactors Due to Decrease of Physical Size	6
1.2.4	Advantages of Microreactors Due to Increase of Number of Units	8
1.3	Potential Benefits of Microreactors Regarding Applications	10
1.4	References	12
2	**Modern Microfabrication Techniques for Microreactors**	15
2.1	Microfabrication Techniques Suitable for Microreactor Realization	15
2.2	Evaluation of Suitability of a Technique	16
2.3	Anisotropic Wet Etching of Silicon	17
2.4	Dry Etching of Silicon	19
2.5	LIGA Process	20
2.6	Injection Molding	21
2.7	Wet Chemical Etching of Glass	22
2.8	Advanced Mechanical Techniques	23
2.8.1	Surface Cutting with Diamond Tools	23
2.8.2	Milling, Turning and Drilling	23
2.8.3	Punching	24
2.8.4	Embossing	24
2.9	Isotropic Wet Chemical Etching of Metal Foils	25
2.10	Electro Discharge Machining (EDM) of Conductive Materials	26
2.10.1	Wire-Cut Erosion and Die Sinking	27
2.10.2	μ-EDM Drilling	29
2.11	Laser Micromachining	29
2.12	Interconnection Techniques	30
2.12.1	Microlamination of Thin Metal Sheets	30
2.13	Functional Coatings	33
2.13.1	Functional Coatings for Corrosion Prevention	33
2.13.2	Functional Coatings for Fouling Prevention	34
2.14	References	35

3	**Micromixers**	41
3.1	Mixing Principles and Classes of Macroscopic Mixing Equipment	41
3.2	Mixing Principles and Classes of Miniaturized Mixers	43
3.3	Potential of Miniaturized Mixers	46
3.4	Contacting of Two Substreams, e.g. in a Mixing Tee Configuration	49
3.4.1	Mixing Tee-Type Configuration	49
3.4.2	Double Mixing Tee-Type Configuration	50
3.5	Collision of High-Energy Substreams for Spraying/Atomizing	52
3.5.1	Collision of Three Substreams in a Microjet Reactor	52
3.6	Injection of Many Small Substreams of One Component into a Main Stream of Another Component	53
3.6.1	Injection of Multiple Microjets	53
3.7	Manifold Splitting and Recombination of a Stream Consisting of Two Fluid Lamellae of Both Components	55
3.7.1	Multiple Flow Splitting and Recombination Combined with Channel Reshaping	55
3.7.2	Multiple Flow Splitting and Recombination Using Fork-Like Elements	58
3.7.3	Multiple Flow Splitting and Recombination Using a Separation Plate	60
3.7.4	Multiple Flow Splitting and Recombination Using a Ramp-Like Channel Architecture	62
3.8	Injection of Many Substreams of Both Components	64
3.8.1	Multilamination of Fluid Layers in an Interdigital Channel Configuration	64
3.8.2	Vertical Multilamination of Fluid Layers Using a V-type Nozzle Array	73
3.8.3	Multilamination Using a Stack of Platelets with Microchannels	75
3.8.4	Multilamination Using a Stack of Platelets with Star-Shaped Openings	79
3.9	Decrease of Diffusion Path Perpendicular to the Flow Direction by Increase of Flow Velocity	80
3.9.1	Decrease of Layer Thickness by Hydrodynamic Focusing	80
3.10	Externally Forced Mass Transport, e.g. by Stirring, Ultrasonic Wave, Electrical and Thermal Energy	83
3.10.1	Dynamic Micromixer Using Magnetic Beads	83
3.11	References	83
4	**Micro Heat Exchangers**	87
4.1	Micro Heat Exchangers with Wide and Flat Channels	89
4.1.1	Cross-Flow Heat Exchange in Stacked Plate Devices	89
4.1.2	Cross-Flow Heat Exchange Based on Cross-Mixing	92
4.1.3	Counter-Flow Heat Exchange in Stacked Plate Devices	94
4.1.4	Electrically Heated Stacked Plate Devices	97
4.2	Micro Heat Exchangers with Narrow and Deep Channels	99
4.2.1	Heat Exchanger with One-Sided Structured Channels	99
4.2.2	Heat Exchanger with Double-Sided Structured Channels	100

4.3	Micro Heat Exchangers with Breakthrough Channels	102
4.4	Axial Heat Conduction	104
4.4.1	Numerical Calculations of the Influence of Material Choice on Heat Transfer Efficiency	104
4.4.2	The Use of Thermal Blocking Structures	105
4.5	Permanent Generation of Entrance Flow by Fins	106
4.6	Generation of a Periodic Flow Profile by Sine-Wave Microchannels	107
4.7	Microtechnology-Based Chemical Heat Pumps	108
4.8	Performance Characterization of Micro Heat Exchangers	109
4.8.1	Temperature Profiles of Micro Heat Exchangers Yielded by Thermograms of Infrared Cameras	110
4.9	References	112

5	**Microseparation Systems and Specific Analytical Modules for Microreactors**	115
5.1	Microextractors	115
5.1.1	Partially Overlapping Channels	115
5.1.2	Wedge-Shaped Flow Contactor	119
5.1.3	Contactor Microchannels Separated by a Micromachined Membrane	122
5.1.4	Contactor Microchannels Separated by Sieve-Like Walls	126
5.1.5	Micromixer—Settler Systems	126
5.2	Microfilters	130
5.2.1	Isoporous-Sieve Microfilters	131
5.2.2	Cross-Flow Microfilters	131
5.3	Gas Purification Microsystems	133
5.4	Gas Separation Microdevices	134
5.5	Specific Analytical Modules for Microreactors	136
5.5.1	Analytical Modules for In-Line IR Spectroscopy	136
5.5.2	Analytical Module for Fast Gas Chromatography	136
5.6	References	140

6	**Microsystems for Liquid Phase Reactions**	143
6.1	Types of Liquid Phase Microreactors	144
6.2	Liquid/Liquid Synthesis of a Vitamin Precursor in a Combined Mixer and Heat Exchanger Device	144
6.3	Acrylate Polymerization in Micromixers	151
6.4	Ketone Reduction Using a Grignard Reagent in Micromixers	154
6.5	Laboratory-Scale Organic Chemistry in Micromixer/Tube Reactors	158
6.6	Dushman Reaction Using Hydrodynamic Focusing Micromixers and High-Aspect Ratio Heat Exchangers	162
6.7	Synthesis of Microcrystallites in a Microtechnology-Based Continuous Segmented-Flow Tubular Reactor	164
6.8	Electrochemical Microreactors	166

6.8.1   Synthesis of 4-Methoxybenzaldehyde in a Plate-to-Plate Electrode
         Configuration 166
6.8.2   Scouting Potentiodynamic Operation of Closed Microcells 169
6.9   References 171

7   Microsystems for Gas Phase Reactions 173
7.1   Catalyst Supply for Microreactors 173
7.2   Types of Gas Phase Microreactors 176
7.3   Microchannel Catalyst Structures 177
7.3.1   Flow Distribution in Microchannel Catalyst Reactors 177
7.3.2   Partial Oxidation of Propene to Acrolein 178
7.3.3   Selective Partial Hydrogenation of a Cyclic Triene 180
7.3.4   H$_2$/O$_2$ Reaction 184
7.3.5   Selective Partial Hydrogenation of Benzene 186
7.3.6   Selective Oxidation of 1-Butene to Maleic Anhydride 187
7.3.7   Selective Oxidation of Ethylene to Ethylene Oxide 187
7.3.8   Reactions Utilizing Periodic Operation 188
7.4   Microsystems with Integrated Catalyst Structures and Heat
         Exchanger 193
7.4.1   Oxidative Dehydrogenation of Alcohols 193
7.4.2   Synthesis of Methyl Isocyanate and Various Other Hazardous Gases 197
7.4.3   H$_2$/O$_2$ Reaction in the Explosion Regime 200
7.5   Microsystems with Integrated Catalyst Structures and Mixer 203
7.5.1   Synthesis of Ethylene Oxide 203
7.6   Microsystems with Integrated Catalyst Structures, Heat Exchanger and
         Sensors 209
7.6.1   Oxidation of Ammonia 209
7.6.2   H$_2$/O$_2$ Reaction 214
7.7   Microsystems with Integrated Mixer, Heat Exchanger, Catalyst
         Structures and Sensors 217
7.7.1   HCN Synthesis via the Andrussov Process 217
7.8   References 224

8   Gas/Liquid Microreactors 229
8.1   Gas/Liquid Contacting Principles and Classes of Miniaturized
         Contacting Equipment 229
8.2   Contacting of Two Gas and Liquid Substreams in a Mixing Tee
         Configuration 232
8.2.1   Injection of One Gas and Liquid Substream 232
8.2.2   Injection of Many Gas and Liquid Substreams into One Common
         Channel 233
8.2.3   Injection of Many Gas and Liquid Substreams into One Packed
         Channel 235

8.2.4	Injection of Many Gas Substreams into One Liquid Channel with Catalytic Walls	237
8.2.5	Injection of Many Gas and Liquid Substreams into Multiple Channels	239
8.3	Generation of Thin Films in a Falling Film Microreactor	244
8.4	References	255

9   **Microsystems for Energy Generation**   257

9.1	Microdevices for Vaporization of Liquid Fuels	257
9.2	Microdevices for Conversion of Gaseous Fuels to Syngas by Means of Partial Oxidations	260
9.2.1	Hydrogen Generation by Partial Oxidations	260
9.2.2	Partial Oxidation of Methane in a Stacked Stainless Steel Sheet System	261
9.2.3	Partial Oxidation of Methane in a Microchannel Reactor	263
9.3	Microdevices for Conversion of Gaseous Fuels to Syngas by Means of Steam Reforming	265
9.3.1	Steam Reforming of Methanol in Microstructured Platelets	265
9.4	References	268

10   **Microsystems for Catalyst and Material Screening**   271

10.1	Parallel Screening of Heterogeneous Catalysts in a Microchannel Reactor	271
10.2	Parallel Screening of Heterogeneous Catalysts in Conventional Mini-Scale Reactors	274
10.3	References	276

11   **Methodology for Distributed Production**   277

11.1	The Miniplant Concept	277
11.1.1	Miniplant Concept for HCN Manufacture	278
11.1.2	The Disposable Batch Miniplant	279
11.2	Paradigm Change in Large-Scale Reactor Design Towards Operability and Environmental Aspects Using Miniplants	280
11.3	References	283

Index   285

1 State of the Art of Microreaction Technology

The aim of this chapter is to define the field referred to as microreaction technology, to analyze principal advantages, to comment on these benefits, reviewing current achievements, to document the state of the art of industrial implementation, and finally to outline future developments.

1.1 Definition

1.1.1 Microsystems Termed Microreactor

In accordance with the term "microsystem", which is widely accepted, microreactors usually are defined as miniaturized reaction systems fabricated by using, at least partially, methods of microtechnology and precision engineering. The characteristic dimensions of the internal structures of microreactors like fluid channels typically range from the sub-micrometer to the sub-millimeter range. Some people also prefer the terms nanoreactors or milli-/minireactors for devices with characteristic dimensions at the lower or the upper boundary of this dimensional range. In this book, however, only the term "microreactor" will be used.

1.1.2 Structural Hierarchy of Microreactors

The construction of microdevices generally is performed in a hierarchic manner, i.e. comprising an assembly of units composed of subunits and so forth. This holds particularly for microreactors which are based on an architecture characterized by multiplying unit cells, the so-called concept of numbering-up [1, 2]. In the following it is aimed to commonly define different units which most often are assembled in microreactors.

Definitions with Regard to Structural Hierarchy
The smallest units of a miniaturized continuous flow system are *microstructures*, in the vast majority of cases referring to *channel* structures (see Figure 1-1). Usually parallel channels are combined to an array surrounded by inlet and outlet flow regions, sometimes referred to as headers. A typical single or multiple flow channel configuration of distinct geometric nature is named *element*. A typical example for a mixing element is an interdigital channel configuration. In some cases, elements can consist of chambers too, e.g. carrying additional microstructures such as pores.

A combination of an element, connecting fluid lines and supporting base material, is termed *unit*. For instance, a mixing platelet with an interdigital structure and feed lines is a micromixing unit. In order to increase throughput, units may form a *stack*, e.g. a stack of catalytic platelets in a chamber of a gas phase microreactor. Alternatively, identical devices

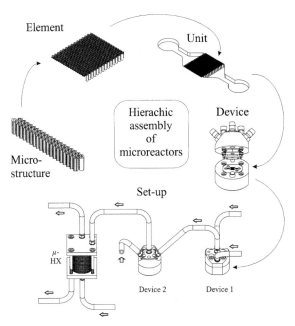

Fig. 1-1. Hierarchical assembly of microreactors, as evidenced for micromixer components.

can be arranged in parallel in a plane, e.g. a micromixer array consisting of thousands of unit cells.

Neither units nor stacks can be operated alone, hence, they are not real microreactors, since they need *housings* or, at least, *top and bottom plates* for fluid connection to external periphery. A *device* refers to a unit embedded either in a housing or between two end caps. The build-up of complex *systems* can be performed by integration of several units within one common housing. A system can also be based on a connection of devices, in this case referred to as *components*.

Any parallel or serial interconnection of components, systems or mixed combinations may be termed *set-up* or *plant*, dependent on the type of application, being lab- or industrial scale oriented, respectively. These set-ups or plants consist of either only microdevices or -systems, or, more likely, may contain microreactors next to conventional larger equipment.

Conceptual Division of Contents of This Book
The structural hierarchy – microstructure/channel, element, unit, device/component, system, set-up – is related to the conceptual division of the contents of this book. Chapters discuss components and systems with respect to different types of reactions and unit operations:

- Micromixers
- Micro heat exchangers
- Microseparators
- Gas phase reactors
- Liquid phase reactors
- Gas/liquid reactors

In the case of a combination of several operations, i.e. referring to a microsystem, the most characteristic function is chosen for classification. For instance, a system consisting of a mixer, heat exchanger and catalyst platelets, designated for carrying out gas phase reactions, will be discussed in the chapter "Gas Phase Reactors". The remaining components of this microsystem will not be described in separate chapters, e.g. in "Micromixers" or "Micro Heat Exchangers", except if they represent a unique flow configuration regarding their microelement. In this case, the presentation of the system is split into a chapter for the (unit) operation and a chapter for the system (reactor).

Very complex assembled microsystems including several types of reactors are discussed in separate chapters. These microreaction systems are referred to their respective type of application. Currently, two special fields of applications are of extremely high commercial interest, namely catalyst/material screening and energy generation. For these applications, two separate chapters were planned in this book. In addition, the assembly of several microsystems into a complete plant, e.g. for distributed production, is discussed in the Chapter 11 "The Miniplant Concept".

In the case of the description of components, sections refer to selected flow configurations, i.e. microelements, typical for a certain function. Since these elements are directly correlated to principles of function, e.g. a certain mixing concept, it was aimed to present a comprehensive overview of present approaches. Thereby, potential advantages of miniaturization are given practical application. Hence, a performance comparison of the various concepts, i.e. components or systems with specific elements, is crucial for a deep understanding of microreaction technology.

For instance, the chapter on micromixers is composed of sections, referring to the microelements termed with respect to the specific flow configuration. To illustrate this type of classification, the following examples of micromixing elements are given:

- Contacting of two substreams, e.g. in a mixing tee configuration
- Collision of two substreams of high energy and generation of a large contact surface due to spraying/atomizing
- Manifold splitting and recombination of a stream consisting of two fluid lamellae of two components

Subsections correspond to specific variants or adaptations of one concept, merely being examples which show the range of possibilities to realize a common idea. For instance, the section "Manifold Splitting and Recombination of a Stream Consisting of Two Fluid Lamellae of Both Components" is subdivided into the following subsections:

- Multiple Flow Splitting and Recombination Combined with Channel Reshaping
- Multiple Flow Splitting and Recombination Using Fork-like Elements
- Multiple Flow Splitting and Recombination Using a Separation Plate etc.

In the case of microreaction systems, this type of classification has not been followed, for reasons listed therein. For instance, in the chapter "Gas Phase Reactors" the microsystems were grouped according to the type of reactions carried out.

1.1.3 Functional Classification of Microreactors

Two classes of microreactors exist, referring to applications in analysis, especially in the field of biochemistry and biology, or chemical engineering and chemistry. Although these fields are distinctly different in most cases, as analytical and preparative equipment are, some microreactors cover both aspects. This holds particularly for combinatorial chemistry and screening microdevices which serve as analytical tools for information gathering as well as synthetic tools providing milligram quantities of products.

A further classification of microreactors is based on the operation mode, either being continuous flow or batch-type. The vast majority of microdevices presented in this book refer to continuous flow systems. Instead, batch systems such as micro and nano titer plates, e.g. for solid-supported chemical synthesis of drugs, will not be reported. In this field, the reader is referred to comprehensive overviews supplied by a number of excellent reviews and books [3–5].

The same holds for a large number of continuous flow microfluidic devices which were developed for analytical purposes starting in the late 1980s. If a series of processes such as filtration, mixing, separation and analysis is combined within one unit, the corresponding microsystems usually were termed micro total analysis systems (µTAS) [6–12]. Most often, these systems were applied for biochemical and chemical analysis. Modern developments consider e.g. polymerase chain reaction, electrophoretic separation, or proteome analysis, just to mention a few [13–15].

Concerning this field already comprehensively described [16], only component development will be presented in the framework of this book which turned out to be relevant for purposes of chemical microreactors. This is especially the case for analytical micromixers yielding a conceptual base for similar constructions for synthetic applications. Hence, in terms of consistency and novelty, the following chapters within this book will only refer to developments concerned with flow-through chemical microreactors, used for process development, production or screening.

1.1.4 Dividing Line Between Analysis and Reaction Systems

Reaction systems generally differ from analysis systems by producing or converting materials or substances. The latter devices are designed to gather information, e.g. to measure the content of a certain analyte in a water sample taken from a lake. However, comparing extremely small individual systems, this difference apparently vanishes, because miniaturization of reaction devices ultimately will decrease the amount of converted materials to a level close to that of analytical devices. Therefore, the productivity of such small reaction devices, which is not sufficient anymore for synthesis purposes, can be used for process development or for screening only.

Both applications clearly refer to measuring tasks, the former regarding the finding of optimum process conditions, the latter of application-tailored materials. Actually, such measuring tools gather information similar to analytical devices. However, the purpose of using the information is distinctly different. In analytics, information gathering is an end in itself. For instance, the detection of ozone concentration in a certain layer of the atmosphere provides important information for ecological research. In contrast, information obtained in small reaction systems is used to optimize a lab synthesis or a large-scale process as well as to produce a new material with advanced properties, thus, finally is related to production issues. Hence, there is a clear dividing line between miniaturized analysis systems and microreactors for chemical applications.

1.2 Fundamental Advantages of Microreactors

Before analyzing the fundamental advantages of microreactors it is worthwhile to shortly review the benefits of miniaturized analysis systems, designed with similar characteristic dimensions as microreactors, and nano-scale reactors of much smaller size.

1.2.1 Fundamental Advantages of Miniaturized Analysis Systems

A large number of applications within the last decade clearly demonstrated fundamental advantages for miniaturized analysis systems compared to lab-scale equipment (see also Section 1.1.3). The smaller devices needed less space, materials, and energy and often had shorter response times [7]. In particular, more information per space and time is gained. By parallel microfabrication and automated assembly, the costs per device could be kept low. Decreasing the component size, in addition, allowed the integration of a multitude of small functional elements, thereby enhancing the system performance [7].

1.2.2 Fundamental Advantages of Nano-Scale Reactors

In the following, a nano-scale reactor is defined as any supramolecular assembly which acts as a reaction unit, i.e. being a host providing a small reaction volume, sometimes encasing only one molecule. To mention only a few, supramolecular assemblies such as molecular tweezers, zeolites, micelles, liposomes and Langmuir–Blodgett layers were, among other applications, utilized as small "reaction vessels". Most often, the molecular

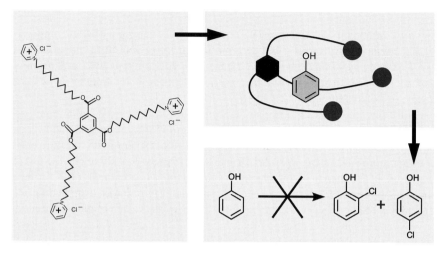

Fig. 1-2. Bola-type amphiphile forming a small nano-sized reaction vessel by means of self-organization. The chlorination of encapsulated phenol is thereby prohibited.

encapsulation strongly modified the reactivity of the reactants, e.g. by electronic interaction of π systems within a bola-type amphiphile [17], by adsorption and isolation in a zeolite cavity [18, 19], or by separation within micelles in order to prevent radical recombination [20].

Hence, small vessels, cavities and clefts provided in nanoreactors allow an interaction by means of molecular forces and modify the electronic structure of reactants. In addition, steric interactions are possible, e.g. influencing the conformation of a molecule or the free rotation of a group attached to a molecule. All these factors, known to modify the reactivity of "free" molecules as well, have a similar effect on "encapsulated" reactants. In this sense, nanoreactors behave like a solvent or a weak complexing agent.

To summarize, the most remarkable feature of nanoreactors is that they are actively changing chemistry, although the encasement certainly influences transport properties as well.

1.2.3 Advantages of Microreactors Due to Decrease of Physical Size

The volumes of microreactors are too large in order to interact with reactants significantly on a molecular level. Their main impact focuses on intensifying mass and heat transport as well as improving flow patterns. Therefore, benefits concerning chemical engineering is the main driver for microreactor investigations, while chemistry, in terms of reaction mechanism and kinetics, remains widely unchanged.

Decrease of Linear Dimensions
Decreasing the linear dimensions increases, for a given difference in a physical property, the respective gradient. This refers to properties particularly important for processing in chemical reactors, such as temperature, concentration, density, or pressure. Consequently, the driving forces for heat transfer, mass transport, or diffusional flux per unit volume or unit area increase when using microreactors.

These simple theoretical predictions were evidenced by a number of studies, e.g. concerning heat and mass transport. The majority of today's microreactor/heat exchanger devices contain microchannels with typical widths of 50 μm to 500 μm; the separating wall material between reaction and heat transfer channels can kept down to 20 to 50 μm, if necessary. As a result, heat transfer coefficients up to 25,000 W/m^2 K measured in microdevices exceed those of conventional heat exchangers by at least one order of magnitude [21]. Typical fluid layer thicknesses in micromixers can be set to a few tens of micrometers, in special configurations down to the nanometer range. Consequently, mixing times in micromixers amount to milliseconds, in some cases even to nanoseconds [22, 23], which is hardly achievable using stirring equipment or other conventional mixers.

Increase of Surface-to-Volume Ratio
As a consequence of the decrease in fluid layer thickness, the corresponding surface-to-volume ratio of the fluid entity is also notably increased. Specific surfaces of microchannels amount to 10,000 to 50,000 m^2/m^3, whereas typical laboratory and production vessels usually do not exceed 1000 m^2/m^3 and 100 m^2/m^3, respectively. Apart from benefits of heat transfer mentioned above, this increase in specific vessel surface can be utilized, e.g., in catalytic gas phase reactors coated with the active material on the inner walls.

Similar benefits have to be expected for multiphase processes, when at least one of the fluid phases has a layer thickness in the micrometer range. Both estimations by theory and experiments proved that the specific interfaces of such multiphases in microreactors can be set in the range of 5000 to 30,000 m^2/m^3. So far, the highest reported interface was measured using a falling film microreactor, amounting to 25,000 m^2/m^3 [24, 25]. Traditional bubble columns do not exceed a few 100 m^2/m^3; the best modern gas/liquid lab contactors such as impinging jets generate liquid surfaces of about 2000 m^2/m^3 [26]. In some favorable cases, e.g. regarding annular flow in micro bubble columns [25, 27], the corresponding specific interfaces can be set nearly as high as the specific surfaces of microchannels, thus potentially achieving 50,000 m^2/m^3 or even larger values. First measurements indeed showed large specific interfaces for this flow pattern, although not yet reaching the theoretically possible limit [25].

Decrease of Volume
Due to the reduction of the linear dimensions, the volume of microreactors is significantly decreased compared to large-scale reactors, typically amounting to a few μl. This difference becomes even larger when, in combination with reactor miniaturization, a large-scale batch process is replaced by continuous flow operation in microdevices. In case of a metallo-organic reaction, the material hold-up could be decreased from a tank of 6000 l size to a

volume of a few milliliters within five miniaturized mixers [28]. The smaller hold-up increases process safety and, due to shorter residence time, improves selectivity.

1.2.4 Advantages of Microreactors Due to Increase of Number of Units

A characteristic feature of microstructured fluidic devices is the multiple repetition of basic units, either fed separately in screening devices or operated in parallel, using a common feed line, for production purposes.

Fast and Cost-Saving Screening of Materials and Processes
Recently, the application of combinatorial strategies has been more and more extended from drug development in pharmacy [3, 4] to screening of inorganic materials, catalysts and polymers [29–32]. Whilst the former processing route was focused on the use of small batch reaction vessels, so-called micro and nano titer plates, the latter approaches demand a diversification of reactor types. So far, inorganic materials, and in selected cases catalysts, were tested as arrays of spatially separated thin zones consisting of different materials [33, 34]. These zones usually were coated on wafers by means of thin film deposition techniques, in most cases using mask processes in order to locally change material properties, e.g. generating concentration profiles.

However, this straightforward approach is limited, especially regarding catalyst and polymer material research. The need for support porosity in the former case and long reaction times in the latter case are two among several arguments favoring the use of tube-like continuous flow reactors. A combined processing in many small tubes in parallel was already realized in flow-through tube reactors [35]. A further decrease in hydraulic diameter leads to more and more compact microreactor design.

First design concepts for such screening microreactors were a stack providing a frame for insertion of disposable catalyst carrier plates [31] and a sheet consisting of a number of reaction plates, structurally analogous to a titer plate [36]. Apart from increasing the number of samples to be investigated, further benefits in screening microreactors are the rapid and precise change of temperature, concentration and pressure. In particular, the possibility of isothermal operation and the high efficiency of mass transfer provide a sound information base, e.g. allowing to measure intrinsic kinetic properties.

Although currently screening in microreactors is most often performed by parallel processing, the continuous flow operation of microreactors enables rapid serial synthesis as well. One possible strategy can be based on the separation of liquid plugs containing the different samples separated by an immiscible liquid in a miniaturized channel [37, 38].

The generation of many samples, either by parallel or serial means, requires fast analysis. The integration of synthesis and analysis in one device has a number of advantages besides compactness. This serves to speed up analysis times, to save sample material and to eliminate additional sample transfer, either performed manually or by a robot.

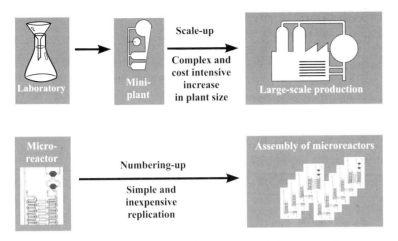

Fig. 1-3. Simplified scheme illustrating scale-up versus numbering-up strategies.

Due to this interdependence of sample generation and testing, screening is, among all application fields of microreactors, strongly dependent on the development of a practical total system approach, combining all relevant components in one device.

Production Flexibility
An increase in throughput in microreactors is achieved by a numbering-up approach, rather than by scaling-up [1, 2]. The functional unit of a microreactor, e.g. a mixing structure, is multiply repeated. Fluid connection between these units can be achieved by using distribution lines and flow equipartition zones, most likely hierarchically assembled (see also Section 1.1.2).

Numbering-up widely guarantees that desired features of a basic unit are kept when increasing the total system size [1, 2]. In an ideal case, measuring devices for process development and production reactor become similar, being composed of identical units. This was, e.g., demonstrated by comparison of mixing quality of an array of ten mixing units with the performance of a single unit [39]. However, these results also showed the crucial importance of flow distribution for the efficiency of the total reaction unit.

In addition, a larger number of units results in higher flexibility in adapting production rate to varying demand since, at least in principle, a certain number of systems can be switched off or further systems may be simply added to the production plant. A plant design based on a large number of small reaction systems can, again at least in principle, be modified to perform a variety of reactions by changing the piping network, i.e. the plant may be adapted to the synthesis of various substances using microreactor modules like a "LEGO" system. This flexibility may be supported by a considerably broader range of operating conditions of a microreactor compared with a macroscopic system.

9

1.3   Potential Benefits of Microreactors Regarding Applications

According to the discussion of the fundamental advantages, the following potential benefits regarding applications of microreactors obtain:

- Faster transfer of research results into production
- Earlier start of production at lower costs
- Easier scale-up of production capacity
- Smaller plant size for distributed production
- Lower costs for transportation, materials and energy
- More flexible response to market demands

Microreactors principally enable a faster transfer of research results into production due to their advantageous setting of operating conditions (see discussion above) yielding more precise data as well as information otherwise not accessible, e.g. regarding new process regimes. Meanwhile, this prediction has been confirmed by a number of examples of use [28, 40–42]. In addition, simple practical aspects make the use of microreactors as measuring tools attractive, e.g. ease and flexibility of construction and disassembly of experimental set-up, small material consumption, or faster approval due to enhanced safety.

Besides research transfer, the other potential applications presented above refer to production issues. So far, the corresponding technological development of respective microreactors could not reach a similar level as for measuring tasks. However, feasibility studies, mainly in the framework of bilateral industrial co-operations, are underway and soon will give a deeper insight [24, 43, 44]. The next steps in these developments will be the realization and testing of production microreactors. Currently, hydrogen production in miniaturized reformers, for subsequent use in fuel cells, seems to be one of the most promising fields that needs light-weight compact microreactors.

Facing production issues, throughput is directly correlated to reaction volume, whereas information gathering is independent of the size of the analysis system. Hence, it is usually assumed that specific production costs increase with decreasing reactor volume, referred to as economy of scale [45]. In addition, the ratio of construction material to reaction volume is inevitably high for microreactors, composed of a multitude of small channels separated by wall material. For similar performance of conventional and microreactors, this unfavorable ratio further seems to render production in microreactors unprofitable. However, specific fields have been identified where either the performance of microreactors is enhanced or the lower specific capacity is counterbalanced by cost saving due to other factors. Some of these fields refer to:

- Replacing a batch by a continuous process
- Intensification of processing
- Safety issues
- Change of product properties
- Distributed production

Replacing a Batch by a Continuous Process
At present, a number of processes are carried out batch-wise, e.g. utilizing stirred vessels. In some cases, reaction time is set much longer than kinetically needed caused by the slow mass and heat transfer in a system of low specific surface area. Replacing this equipment by a continuous flow process in a microreactor can, due to fast transport in thin fluid layers, result in notably decreased contact times. In total, the process may be carried out faster. In addition, conversion and selectivity may increase as demonstrated for metallo-organic reactions [24, 28]. Hence, space–time yields of such microreactors can exceed that of batch processes.

Intensification of Processing
Due to short diffusional distances, conversion rates can be significantly enhanced in microsystems. This was exemplarily verified for the intensification of heat transfer of thermally coupled high-temperature reactions by recent calculations [44]. For a given chemical process, using conventional technology and a microreactor, calculations predict that the amount of catalyst needed can be decreased by miniaturization by nearly a factor of 1000. This increase in performance was gained at the expense of a decrease of the ratio of reactor volume to construction material in case of the microreactor. Nevertheless, the microreactor size could be decreased by a factor of 10 compared with conventional technology, rendering its performance superior for this specific application.

Safety Issues
In the past, several examples of processing in microreactors clearly demonstrated safe operation using process parameters of otherwise explosive regimes [46–48]. This information cannot be used for classical process development, aiming at an increase of reactor size. The production inevitably has to be achieved by numbering-up of identical units, i.e. keeping the individual reaction units small. The use of operating conditions, corresponding to so far "remote" or "secret" regimes, certainly expands the operational flexibility of conventional reactors. Currently, the importance of this expansion cannot put in concrete terms, if being more restricted to model reactions, e.g. contacting hydrogen and oxygen, or being a general concept for several examples of use.

Change of Product Properties
So far, only the influence of transport intensification on conversion rate, conversion and selectivity was discussed. This holds as far as the synthesis of low-molecular weight molecules is concerned. Taking into account polymer and multiphase formation, conformational and compositional features become important as well as morphological properties, e.g. as given for the generation of supramolecular structures such as liposomes, capsules and microemulsions. These features are sensitive particularly to micromixing phenomena designating microsystems for application in these fields. Highly efficient and homogeneous mass transfer in micromixers was demonstrated to improve the uniformity of the weight distribution of polymers [41] and size distribution of droplets in semi-solid pastes [49].

Distributed Production

At present, production is carried out in large plants and it is aimed to make them larger as far as technically feasible. Concerning petrochemistry, for example, the contents of natural feedstocks are transported, sometimes over long distances, to a central plant and converted therein to more valuable products. The small size and remote location of a vast number of feedstocks render exploitation not profitable since neither plant construction nor transport in pipelines is economically attractive. Processing in microreactors may be less costly than applying conventional equipment. Installation and removal may be sufficiently fast and flexibility towards productivity high due to the numbering-up assembly.

Other examples of processes which potentially could benefit from a distributed production were given by many authors [2, 45, 50]. Most often, reactions with extremely toxic substances or otherwise hazardous potential were mentioned.

These potential features illustrate that microreactors are promising tools for on-site and on-demand production.

1.4 References

[1]   Ehrfeld, W., Hessel, V., Haverkamp, V.; *"Microreactors"*, *Ullmann's Encyclopedia of Industrial Chemistry*, Wiley-VCH, Weinheim, (1999).
[2]   Lerou, J. J., Harold, M. P., Ryley, J., Ashmead, J., O'Brien, T. C., Johnson, M., Perrotto, J., Blaisdell, C. T., Rensi, T. A., Nyquist, J.; *"Microfabricated mini-chemical systems: technical feasibility"*, in Ehrfeld, W. (Ed.) *Microsystem Technology for Chemical and Biological Microreactors*, Vol. 132, pp. 51–69, Verlag Chemie, Weinheim, (1996).
[3]   Weber, L.; *"Kombinatorische Chemie – Revolution in der Pharmaforschung?"*, Nachr. Chem. Tech. Lab. **42**, pp. 698–702, 7 (1994).
[4]   Balkenhohl, F., von dem Bussche-Hünnefeld, C., Lansky, A., Zechel, C.; *"Kombinatorische Synthese niedermolekularer organischer Verbindungen"*, Angew. Chem. **108**, pp. 2437–2488 (1996).
[5]   Jung, G., Beck-Sickinger, A. G.; *"Methoden der multiplen Peptidsynthese und ihre Anwendungen"*, Angew. Chem. **104**, pp. 375–500, 4 (1992).
[6]   Harrison, D. J., Fluri, K., Seiler, K., Fan, Z., Effenhauser, C. S., Manz, A.; Science **261**, p. 895 (1993).
[7]   Widmer, H. M.; *"A survey of the trends in analytical chemistry over the last twenty years, emphasizing the development of TAS and µTAS"*, in Widmer, E., Verpoorte, E., Banard, S. (Eds.) *Proceedings of the 2nd International Symposium on Miniaturized Total Analysis Systems, µTAS96 – Special Issue of Analytical Methods & Instrumentation AMI*, pp. 3–8, Basel, (1996).
[8]   van den Berg, A., Bergveld, P.; *"Development of µTAS concepts at the MESA Research institute"*, in Widmer, E., Verpoorte, E., Banard, S. (Eds.) *2nd International Symposium on Miniaturized Total Analysis Systems, µTAS96 – Special Issue of Analytical Methods & Instrumentation AMI*, pp. 9–15, Basel, (1996).
[9]   Ramsey, J. M.; *"Miniature chemical measurement systems"*, in Widmer, E., Verpoorte, E., Banard, S. (Eds.) *Proceedings of the 2nd International Symposium on Miniaturized Total Analysis Systems, µTAS96 – Special Issue of Analytical Methods & Instrumentation AMI*, pp. 24–27, Basel, (1996).
[10]   Northrup, M. A., Benett, B., Hadley, D., Stratton, P., Landre, P.; *"Advantages afforded by miniaturization and integration of DNA analysis instrumentation"*, in Ehrfeld, W. (Ed.) *Microreaction Technology, Proceedings of the 1st International Conference on Microreaction Technology; IMRET 1*, pp. 278–288, Springer-Verlag, Berlin, (1997).
[11]   Woolley, A. T., Mathies, R. A.; *"Ultra-high-speed DNA fragment separations using microfabricated capillary array electrophoresis chip"*, Proc. Natl. Acad. Sci. USA **91**, (1994) 11348–11352.
[12]   van der Schoot, B. H., Verpoorte, E. M. J., Jeanneret, S., Manz, A., de Rooji, N. F.; *"Microsystems for analysis in flowing solutions"*, in Proceedings of the "Micro Total Analysis Systems, µTAS '94", 1994; pp. 181–190; Twente, Netherlands.

[13] Manz, A.; *"The secret behind electrophoresis microstructure design"*, in Widmer, E., Verpoorte, E., Banard, S. (Eds.) *Proceedings of the 2nd International Symposium on Miniaturized Total Analysis Systems, μTAS96 – Special Issue of Analytical Methods & Instrumentation AMI*, pp. 28–30, Basel, (1996).

[14] Köhler, J. M., Dillner, U., Mokansky, A., Poser, S., Schulz, T.; *"Micro channel reactors for fast thermocycling"*, in Ehrfeld, W., Rinard, I. H., Wegeng, R. S. (Eds.) *Process Miniaturization: 2nd International Conference on Microreaction Technology; Topical Conference Preprints*, pp. 241–247, AIChE, New Orleans, USA, (1998).

[15] Lottspeich, F.; *"Proteomanalyse – ein Weg zur Funktionsanalyse von Proteinen"*, Angew. Chem. **111,** 17 (1999) 2630–2647.

[16] Manz, A., Becker, H.; *"Microsystem Technology in Chemistry and Life Science"*, Topics in Current Chemistry, Vol. 194, Springer-Verlag, Heidelberg, (1998).

[17] Suckling, C. J.; *"Host-guests binding by a simple detergent derivative: tentacle molecules"*, J. Chem. Soc., Chem. Commun., pp. 661–662 (1982).

[18] Barrer, B.; *"Chemical nomenclature and formulation of compositions of synthetic and natural zeolites"*, Pure Appl. Chem. **51,** pp. 1091–1100 (1979).

[19] Drzaj, R.; *"Zeolites. synthesis, structure, technology and application"*, in Proceedings of the "Int. Symp. on Synthesis of Zeolites, their Structure Determination and their Technological Use", 1985; Ljubljana.

[20] Fikentscher, H., Gerrns, H., Schuller, H.; Angew. Chem. **72,** pp. 856–864 (1960).

[21] Schubert, K., Bier, W., Brandner, J., Fichtner, M., Franz, C., Linder, G.; *"Realization and testing of microstructure reactors, micro heat exchangers and micromixers for industrial applications in chemical engineering"*, in Ehrfeld, W., Rinard, I. H., Wegeng, R. S. (Eds.) *Process Miniaturization: 2nd International Conference on Microreaction Technology, IMRET 2; Topical Conference Preprints*, pp. 88–95, AIChE, New Orleans, USA, (1998).

[22] Branebjerg, J., Gravesen, P., Krog, J. P., Nielsen, C. R.; *"Fast mixing by lamination"*, in Proceedings of the "IEEE-MEMS '96", 12–15 Febr., 1996; pp. 441–446; San Diego, CA.

[23] Knight, J. B., Vishwanath, A., Brody, J. P., Austin, R. H.; *"Hydrodynamic focussing on a silicon chip: mixing nanoliters in microseconds"*, Phys. Rev. Lett. **80,** p. 3863, 17 (1998).

[24] Hessel, V., Ehrfeld, W., Golbig, K., Haverkamp, V., Löwe, H., Storz, M., Wille, C., Guber, A., Jähnisch, K., Baerns, M.; *"Gas/liquid microreactors for direct fluorination of aromatic compounds using elemental fluorine"*, in Ehrfeld, W. (Ed.) *Microreaction Technology: 3rd International Conference on Microreaction Technology, Proceedings of IMRET 3*, pp. 526–540, Springer-Verlag, Berlin, (2000).

[25] Hessel, V., Ehrfeld, W., Herweck, T., Haverkamp, V., Löwe, H., Schiewe, J., Wille, C., Kern, T., Lutz, N.; *"Gas/liquid microreactors: hydrodynamics and mass transfer"*, in Proceedings of the "4th International Conference on Microreaction Technology, IMRET 4", pp. 174–187; 5–9 March, 2000; Atlanta, USA.

[26] Herskowits, D., Herskowits, V., Stephan, K.; *"Characterization of a two-phase impinging jet absorber – II. absorption with chemical reaction of CO_2 in NaOH solutions"*, Chem. Engin. Sci. **45,** pp. 1281–1287 (1990).

[27] Chambers, R. D., Spink, R. C. H.; *"Microreactors for elemental fluorine"*, Chem. Commun., pp. 883–884 (1999).

[28] Krummradt, H., Kopp, U., Stoldt, J.; *"Experiences with the use of microreactors in organic synthesis"*, in Ehrfeld, W. (Ed.) *Microreaction Technology: 3rd International Conference on Microreaction Technology, Proceedings of IMRET 3*, pp. 181–186, Springer-Verlag, Berlin, (2000).

[29] Cong, P., Doolen, R. D., Fan, Q., Giaquinta, D. M., Guan, S., McFarland, E. W., Poojary, D. M., Self, K., Turner, H. W., Weinberg, W. H.; *"Kombinatorische Parallelsynthese und Hochgeschwindigkeitsrasterung von Heterogenkatalysator-Bibliotheken"*, Angew. Chem. **111,** p. 508, 4 (1999).

[30] Jandeleit, B., Schaefer, D. J., Powers, T. S., Turner, H. W., Weinberg, W. H.; *"Kombinatorische Materialforschung und Katalyse"*, Angew. Chem. **111,** p. 2649, 17 (1999).

[31] Zech, T., Hönicke, D., Lohf, A., Golbig, K., Richter, T.; *"Simutaneous screening of catalysts in microchannels: methodology and experimental setup"*, in Ehrfeld, W. (Ed.) *Microreaction Technology: 3rd International Conference on Microreaction Technology, Proceedings of IMRET 3*, pp. 260–266, Springer-Verlag, Berlin, (2000).

[32] Rodemerck, U., Ignaszewski, P., Lucas, M., Claus, P., Baerns, M.; *"Parallel synthesis and testing of heterogeneous catalysts"*, in Ehrfeld, W. (Ed.) *Microreaction Technology: 3rd International Conference on Microreaction Technology, Proceedings of IMRET 3*, pp. 287–293, Springer-Verlag, Berlin, (2000).

[33] Danielson, E., Golden, J. H., McFarland, E. W., Reaves, C. M., Weinberg, W. H., Wu, X. D.; Nature **389,** pp. 944–948 (1997).

13

[34] Gong, P., Doolen, R. D., Fan, Q., Giaquinta, D. M., Guan, S., McFarland, E. W., Poojary, D. M., Self, K., Turner, H. W., Weinberg, W. H.; Angew. Chem. (Int. Ed.) **38**, pp. 484–488 (1999).

[35] Senkan, S. M., Ozturk, S.; *"Entdeckung und Optimierung von Heterogenkatalysatoren durch kombinatorische Chemie"*, Angew. Chem. **111**, p. 867, 6 (1999).

[36] Results of IMM, unpublished

[37] de Bellefon, C., Tanchoux, N., Caraveilhes, S., Hessel, V.; *"New reactors for liquid-liquid catalysis"*, in Proceedings of the "Entretiens Jaques Cartier", 6–8 Dec., 1999; Lyon.

[38] de Bellefon, C., Caraveilhes, S., Joly-Vuillemin, C., Schweich, D., Berthod, A.; *"A liquid-liquid plug flow continuous reactor for the investigation of catalytic reactions: the centrifugal partition chromatograph"*, Chem. Eng. Sci. **53**, p. 71–74, 1 (1998).

[39] Ehrfeld, W., Golbig, K., Hessel, V., Löwe, H., Richter, T.; *"Characterization of mixing in micromixers by a test reaction: single mixing units and mixer arrays"*, Ind. Eng. Chem. Res. **38**, pp. 1075–182, 3 (1999).

[40] Wörz, O., Jäckel, K. P., Richter, T., Wolf, A.; *"Microreactors, new efficient tools for optimum reactor design"*, Microtechnologies and Miniaturization, Tools, Techniques and Novel Applications for the BioPharmaceutical Industry, IBC's 2nd Annual Conference; Microtechnologies and Miniaturization, Frankfurt, Germany, (1998).

[41] Bayer, T., Pysall, D., Wachsen, O.; *"Micro mixing effects in continuous radical polymerization"*, in Ehrfeld, W. (Ed.) *Microreaction Technology: 3rd International Conference on Microreaction Technology, Proceedings of IMRET 3,* pp. 165–170, Springer-Verlag, Berlin, (2000).

[42] Srinivasan, R., I-Ming Hsing, Berger, P., Jensen, E. K. F., Firebaugh, S. L., Schmidt, M. A., Harold, M. P., Lerou, J. J., Ryley, J. F.; *"Micromachined reactors for catalytic partial oxidation reactions"*, AIChE J. **43**, pp. 3059–3069, 11 (1997).

[43] Hessel, V., Ehrfeld, W., Golbig, K., Hofmann, C., Jungwirth, S., Löwe, H., Richter, T., Storz, M., Wolf, A., Wörz, O., Breysse, J.; *"High temperature HCN generation in an integrated Microreaction system"*, in Ehrfeld, W. (Ed.) *Microreaction Technology: 3rd International Conference on Microreaction Technology, Proceedings of IMRET 3,* pp. 151–164, Springer-Verlag, Berlin, (2000).

[44] Hardt, S., Ehrfeld, W., vanden Bussche, K. M.; *"Strategies for size reduction of microreactors by heat transfer enhancement effects"*, in Proceedings of the "4th International Conference on Microreaction Technology, IMRET 4", 5–9 March, 2000; Atlanta, USA.

[45] Rinard, I. H.; *"Miniplant design methodology"*, in Ehrfeld, W., Rinard, I. H., Wegeng, R. S. (Eds.) *Process Miniaturization: 2nd International Conference on Microreaction Technology; Topical Conference Preprints,* pp. 299–312, AIChE, New Orleans, USA, (1998).

[46] Hagendorf, U., Janicke, M., Schüth, F., Schubert, K., Fichtner, M.; *"A Pt/Al₂O₃ coated microstructured reactor/heat exchanger for the controlled H₂/O₂-reaction in the explosion regime"*, in Ehrfeld, W., Rinard, I. H., Wegeng, R. S. (Eds.) *Process Miniaturization: 2nd International Conference on Microreaction Technology; Topical Conference Preprints,* pp. 81–87, AIChE, New Orleans, USA, (1998).

[47] Veser, G., Friedrich, G., Freygang, M., Zengerle, R.; *"A modular microreactor design for high-temperature catalytic oxidation reactions"*, in Ehrfeld, W. (Ed.) *Microreaction Technology: 3rd International Conference on Microreaction Technology, Proceedings of IMRET 3,* pp. 674–686, Springer-Verlag, Berlin, (2000).

[48] Ehrfeld, W., Hessel, V., Löwe, H.; *"Extending the knowledge base in microfabrication towards themical engineering and fluid dynamic simulation"*, in Proceedings of the "4th International Conference on Microreaction Technology, IMRET 4", 5–9 March, 2000; Atlanta, USA.

[49] Hessel, V., Ehrfeld, W., Haverkamp, V., Löwe, H., Schiewe, H.; *"Dispersion Techniques for Laboratory and Industrial Production"*, in Müller, R. H. (Ed.) Apothekerverlag, Stuttgart, (1999); in press.

[50] Ponton, J. W.; *"Observations on hypothetical miniaturised disposable chemical plant"*, in Ehrfeld, W. (Ed.) *Microreaction Technology, Proceedings of the 1st International Conference on Microreaction Technology; IMRET 1,* pp. 10–19, Springer-Verlag, Berlin, (1997).

2 Modern Microfabrication Techniques for Microreactors

A number of publications exclusively refer to manufacturing aspects of microreactors, highlighting specifically adapted microstructuring and interconnection techniques [1–7]. Broad overviews, covering a large number of fabrication technologies and several examples of use, were reported as well [8–11].

Therefore, in the following sections, the state of the art of existing microfabrication technologies will only briefly be summarized. It is aimed, however, to focus on adaptations or special variants of these techniques which were exclusively developed for microfluidic applications.

2.1 Microfabrication Techniques Suitable for Microreactor Realization

At present, the following technologies have been preferably applied for fabrication of microreactors:

- Bulk micromachining of monocrystalline materials, e.g. silicon, using anisotropic wet chemical etching [12, 13]
- Dry etching processes using low pressure plasma or ion beams (reactive ion etching, reactive ion beam etching) [14, 15]
- A combination of deep lithography, electroforming and molding [16], micromachining with laser radiation (LIGA process) [17, 18]
- Micromolding, e.g. using mold inserts machined by precision engineering techniques [19–21]
- Wet chemical etching of glass including anisotropic etching of photosensitive glass [22]
- Advanced mechanical milling, turning, sawing, embossing, punching, and drilling processes based on precision engineering [23–28]
- Isotropic wet chemical etching, e.g. of metal foils with a resist pattern [29]
- Micro electro discharge machining [30–32]
- Laser ablation [33]

Usually, several technologies are combined in a process line for fabricating microreactors so that these technologies can be regarded like tool machines in a micro workshop. In addition, auxiliary processes ranging from thin film technologies to mechanical surface modification belong to the tool assembly of this micro workshop as well as test equipment and design and simulation software.

2.2   Evaluation of Suitability of a Technique

The variety of microtechniques certainly can be confusing for a customer who desires to use microproducts for his specific application. Even specialists in the field of microfabrication will tend to use technologies with which they are familiar. As shown in the following list and exemplarily in Figure 2-1, certain factors have to be primarily considered by producers when choosing a fabrication technology.

- Process costs
- Process time
- Accuracy
- Reliability
- Material choice
- Access

Hence, it is worthwhile after finishing a conceptual design of a specific microreaction system to compare the different technologies with respect to the above mentioned parameters. But is reduction of costs the ultimate target? This certainly holds for all microreactors designated for production issues. On the other hand, is quality of the information to be

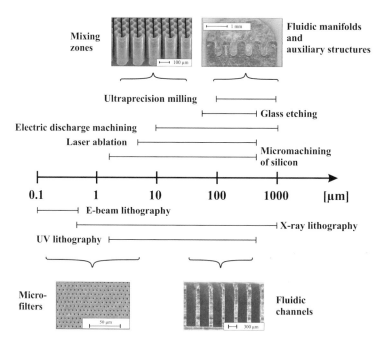

Fig. 2-1. Illustration aiming at a ranking of fabrication technologies with respect to their accuracy and correlation of this figure to typical characteristic dimensions of current microreactor components.

gathered most important? In this case, more expensive fabrication and use of high-value materials may be not criteria of exclusion. In some cases, e.g. regarding safety issues, it may be advisable to build the most reliable system which not may necessarily correspond to the highest-performance version. This will influence the choice of technique as well.

Usually, decisions are made on an overall evaluation of technological potential. For instance, mass production capability of a process refers to a complex function of its costs, time and reliability. The same parameters are of course important for the fabrication of unique specimens as well, but in case of highly profitable applications, high costs sometimes not being prohibitive. Accuracy and material choice of a fabrication technology define the range of applicability of the respective microreactors to a specific chemical process.

Finally, access to a microstructuring technique is a factor not to be underestimated. The high investment costs for processing equipment and maintenance still limit a worldwide widespread distribution of suppliers. In addition, intense communication between supplier and customer about manufacturing capability and needs of an application, respectively, is required.

All these examples, taken from practical experiences of the authors, served to underline the importance of exact clarification and rating of targets before undertaking a long-term development program since probably not all can be achieved to the same extent. Therefore, the following sections aim at shortly introducing different fabrication technologies by explanation of the main processing steps. For a detailed description of the fabrication the reader is referred to the special literature cited as well as reviews and books in this field [34–37].

In addition, the range of applications with respect to material choice and accuracy is illustrated in the sections by presenting selected microfluidic structures realized by the respective technology. This is completed by summarizing comments in order to evaluate the potential of the specific fabrication processes qualitatively, taking into account process time and costs as well as access.

Certainly, the development of fabrication technologies is a rapid, on-going process rendering the following evaluation to be a snapshot of the current needs of microreaction technology. This analysis, moreover, is made from the view of a customer of microreactors, namely a chemist or chemical engineer. It is probably not representative for other application fields of microdevices such as information technology, biotechnology, or sensor technology.

2.3 Anisotropic Wet Etching of Silicon

Wet-chemical etching of single-crystalline silicon (bulk and surface micromachining) was the first process suitable for mass fabrication of micromechanical components [12, 13]. This process relies on a difference of etch velocities with respect to the crystalline directions for a number of etchants. Automation of many processing steps is achievable, rendering this technology attractive for small-series production. A drawback is that the equipment required is relatively costly, taking also into account the need for processing within a clean room. By means of batch processing, several structures can be realized in parallel on one wafer.

A variety of simple geometric structures like grooves, channels, filters, cantilevers or membranes, can be realized by means of patterned etch stop layers that are generated by lithographic and thin film processes on a silicon wafer. Similar structures were employed for a number of microreactor components like pumps, valves, or static mixers which were most often designated for analytical purposes. Silicon micromachining, in particular, benefits from existing bonding processes, either thermally or anodically, which can be utilized for irreversible assembly of microstructured layers [38].

A further advantage of using silicon microfabrication is its ease of combination with thin film technology. Thereby, silicon channel surfaces can be patterned, e.g. yielding catalyst layers, heaters and sensors. This enables a high degree of system integration to be achieved as, e.g., evidenced by a microreaction system for gas phase processes [39, 40].

For example, many micromixers and -extractors were realized as sandwiches composed of silicon and glass layers (see Figure 2-2) [41–46]. This allowed direct visualization of the corresponding processes. Silicon chip technology was also employed for realization of complex microreaction systems designed for carrying out gas phase reactions at high temperatures [47]. The ability to generating thin membranes, in combination with the thermal properties of silicon and thin film deposited materials, proved to be particularly favorable for process control of catalytic reactions [48]. On no account, including liquid phase reactions with aggressive chemicals [49], were limitations of material applicability worth mentioning, e.g. related to corrosion, noticed.

Hence, silicon certainly is a favorable material whenever small sized microstructures of small dimensions and high precision are demanded. Considering platelets of large size, e.g. of several tens of millimeters extension, cost benefits by parallel batch processing are no longer provided, preferring other mass fabrication techniques based on lower technical expenditure, e.g. punching or etching. In addition, if characteristic dimensions become larger, e.g., than 100 μm, the efforts to achieve high precision may no longer be counterbalanced by an improvement with respect to chemical processing.

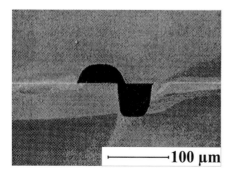

Fig. 2-2. Cross-section of a microextractor consisting of etched silicon and glass layers [46].

Currently, these arguments are only tentative since so far no systematic study has investigated the influence of real channel geometry on final reactor performance. To illustrate this lack of information, it is, e.g., not known if microstructuring techniques yielding smooth surfaces are superior to those providing rough ones, see e.g. [50].

2.4 Dry Etching of Silicon

Highly anisotropic removal of material is obtained by so-called dry etching processes using directed ions from a low-pressure plasma or an ion beam generated at high vacuum [14, 15, 51]. A number of process variants exist mostly based on a variation of the reactive gases used.

Compared with wet chemical etching, a number of advantages can be observed, e.g. fewer geometrical restrictions concerning the microstructures, to a still improved structural resolution and to an extension of material choice. Deep structures with high aspect ratios were produced by reactive ion etching processes (RIE) using low temperature conditions or side wall passivation techniques, a further process variant being termed advanced silicon etching (ASE) [14]. This technology can also be extended to the fabrication of movable structures using so-called sacrificial layers which are removed by additional wet chemical etching.

By means of the ASE technique, two types of silicon micromixers were realized, one thereof being previously fabricated by the LIGA technique. The choice of this technique was especially motivated by the high precision and high aspect ratio required. One of the micromixers was used for contacting highly corrosive solutions which are not compatible with any stainless steel alloy (see Figure 2-3). Thus, the chemical resistance of silicon had to be improved by protection with a thermal oxide layer.

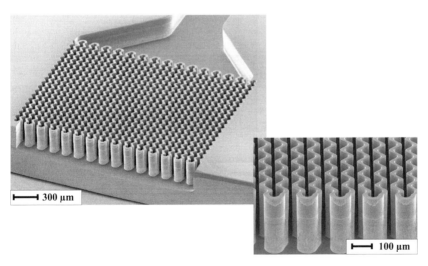

Fig. 2-3. Scanning electron micrograph of an interdigital micromixing element made of silicon realized by advanced silicon etching [52].

2.5 LIGA Process

LIGA and related techniques are based on a sequence of process steps combining (deep) lithography [53–55], electroforming [56] and molding [19, 57, 58] (German acronym: **Li**thographie, **G**alvanoformung, **A**bformung). In a first step, a microstructure is generated by means of a lithographic process in a thick resist layer. This can be achieved by utilizing a laser, a high-energy electron or ion beam as well as standard UV- or X-ray lithography with synchrotron radiation. In the latter case, X-rays allow ultraprecise microstructures with large aspect ratios to be produced. In a second step, a complementary metal structure is generated from the resist master by means of electroforming. The metal structure is used, in a third step, as a mold insert or embossing tool allowing replication processes to be performed such as injection molding or embossing, respectively. Thereby, microstructures preferably made of plastics are mass manufactured. First feasibility tests confirmed that the same holds for ceramic or stainless steel materials.

Fabrication of microstructures using the LIGA process is characterized by high precision, high surface quality and high aspect ratios. Unlike many other processes, which are monolithically based on one material or one class of materials, this process is based on a broad material palette ranging from metals, metal alloys and ceramics to polymers.

A number of parts of microreactor components or systems were achieved as metal structures fabricated by electroforming in the framework of the LIGA process [8, 9, 59]. This selection includes microfilters, micromixers, micro heat exchangers, microextractors, and gas/liquid microreactors (see Figure 2-4).

By means of injection molding, polymeric structures for micropumps [8, 60] and micromixers [8, 18] were realized as well. Micropumps meanwhile are available as inexpensive demonstrators by small-series fabrication using a semi-automated laser assembly technique. Injection molding further allows one to manufacture ceramic microstructures from feedstocks, consisting of polymer binders and ceramic powders [61–63].

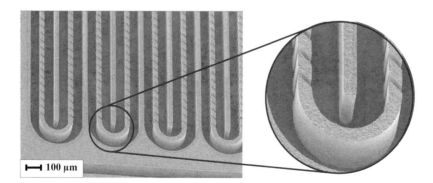

Fig. 2-4. Scanning electron micrograph of a microextractor with perforated walls, made of nickel on copper, for separation of fluids flowing in counter-flow. Realization of this microstructure was achieved by electroforming in the framework of the LIGA process [8].

Concerning a further stage of development of microreaction technology, namely the need for inexpensive mass production, molding processes in the framework of the LIGA technique will gain in importance. However, the unique features of LIGA, combining high precision with fast replication, will have to compete with the technology base provided by other established technologies, e.g. referring to punching or etching.

2.6 Injection Molding

Micromolding has not only been performed in the framework of the LIGA process [20, 21], but relies as well on mold inserts made by other techniques, e.g. microerosion and precision engineering.

A particular interesting approach, termed hot molding, was developed to simplify existing ceramic molding techniques [61]. This process modification allows one to operate at lower working temperatures and injection pressures. This is achieved by using paraffins and low-viscosity waxes with low melting temperatures instead of using standard binders. Further advantages refer to machine and tooling costs, allowing even the use of aluminum or plastics as mold inserts. As models for microfluidic systems, aluminum oxide plates with straight channels or meanders were fabricated by means of the hot molding technique (see Figure 2-5).

Another example of use refers to the manufacture of miniaturized electrical heaters within microchannel structures by injection molding. Mixtures of Al_2O_3/TiN turned out to be suitable materials for this purpose [64], in particular due to the favorable ratio of mechanical stress for a wide variety of compositions of this mixture. In feasibility experiments temperatures up to 1350 °C were achieved.

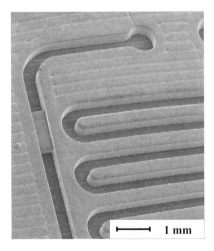

1 mm

Fig. 2-5. Scanning electron micrograph of a meander pattern generated by hot molding of an aluminum oxide-wax feedstock [61].

21

It has to be mentioned that, due to the use of polymeric binders, ceramic microprocessing techniques are in some cases limited regarding their application. Inherent problems of these techniques are shrinkage during the sintering process and warpage effects. In addition, the surface roughness can usually not be set below a few microns due to the relatively large grain size of ceramic particles.

2.7 Wet Chemical Etching of Glass

Microstructuring of glass can be performed by standard lithographic methods [65] or by a particular variant of this process using wafers made of a photosensitive glass (FOTURAN®) [22]. In the latter case, glass wafers are exposed to UV light through a quartz mask with a chromium absorber pattern, well known from semiconductor processing. Thereby, silver nuclei are generated in the glass matrix by a photochemical process. Crystallization of the exposed parts occurs by subsequent annealing which thereafter are removed by etching with hydrofluoric acid.

A number of microdevices made of glass, preferably mixers, extractors, and other separation components, were manufactured by means of etching technologies. Most often, a hybrid solution was chosen, connecting the microstructured glass layer to a silicon wafer. In addition, single micromixers (see Figure 2-6) and micro heat exchangers as well as complete parallel operating microsystems consisting of mixers and heat exchangers were realized in FOTURAN® and tested [66].

Major advantages of using glass in chemical and biochemical microreactors are its material properties such as high chemical and temperature resistance and biocompatibility. Compared with silicon micromaching, structural precision is certainly limited and assembly techniques are presently not provided to a similar high standard. First components for biotechnical applications using microstructured FOTURAN® glass were reported recently [67].

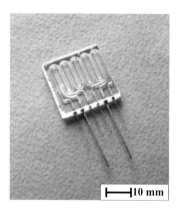

Fig. 2-6. Photograph (left) and scanning electron micrograph (right) of a micro heat exchanger made of FOTURAN® realized by wet chemical etching (by courtesy of mgt mikroglas technik AG).

2.8 Advanced Mechanical Techniques

Owing to the fact that stainless steel is a preferred material for chemical applications, advanced precision engineering techniques gained considerable importance for fabrication of microreactors. Existing irreversible interconnection techniques such as diffusion bonding and laser welding have been adapted to the needs of microfluidic structures. A large variety of techniques have been applied for the realization of platelets carrying microfluidic structures, including milling [25], drilling [27], punching, and embossing [28].

2.8.1 Surface Cutting with Diamond Tools

For microstructuring and interconnection of thin foils to a metal stack, a combination of an advanced milling technique, referred to as surface shaping, with diffusion and electron beam welding was systematically studied in particular [26]. Parallel grooves of usually rectangular shape are made by ground-in precision diamonds or ceramic inserts as cutting tools. Meanwhile, micro heat exchangers, micromixers and microreactors are produced in small series using foil surface shaping (see Figure 2-7).

2.8.2 Milling, Turning and Drilling

Milling, turning and drilling are often used for the fabrication of larger fluidic channels in the range of several hundred micrometers. Another major field of application is related to the design of the housing, namely manufacturing of feed and distribution lines as well as chambers for platelet stacks and sealing grooves.

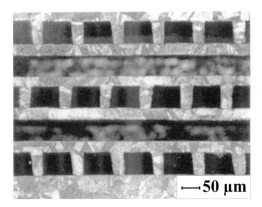

Fig. 2-7. Cut view of a stack of microstructured platelets of a cross-flow heat exchanger realized by surface cutting with diamond tools [68].

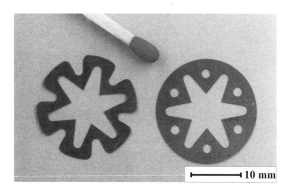

Fig. 2-8. Photograph of two star-shaped micromixing platelets made of stainless steel realized by punching [69].

2.8.3 Punching

The application of punching for realization of microfluidic structures mainly refers to the removal of material within platelets. Thereby, open fluidic structures of micromixers (see Figure 2-8) and micro heat exchangers were generated [69]. This process allows inexpensive and fast structuring of a large number of platelets.

Since the platelet thickness defines the channel depth, i.e. the characteristic dimension of the fluid system, microreactor performance is directly correlated to this parameter. Hence, sufficiently thin platelets, e.g. of a few 10 µm thickness, should be chosen. However, so far no analysis has been made considering deviations of the channel depth by bending or insufficient tightening of platelets. Hence, ease of structuring may compete with precision achieved.

2.8.4 Embossing

Embossing techniques have been especially used for structuring of ceramic green tapes, composed of ceramic powder and polymeric binder material. After tape casting of the ceramic slurry, embossing is performed. This technique was used, e.g., to generate microchannels with diameter of 120 µm in tapes made of Y-ZrO$_2$/Al$_2$O$_3$ and Al$_2$O$_3$/TiN. After sintering, ceramic platelets were yielded and assembled to a micro heat exchanger operating in a cross-flow mode. Figure 2-9 shows a cut through the platelet stack showing cross-sections of the parallel microchannels [61, 70].

Besides ceramic green tapes, embossing has been applied to aluminum and stainless steel platelets. Heat exchanger structures were realized as well [8, 71].

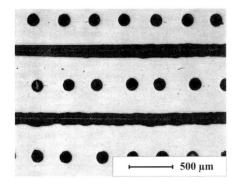

Fig. 2-9. Scanning electron micrograph of a cross-section of a cross-flow heat exchanger made of Y-ZrO$_2$/Al$_2$O$_3$ by embossing [61].

2.9 Isotropic Wet Chemical Etching of Metal Foils

A resist coated on metal foils is patterned by UV irradiation through a mask. Thereafter, the exposed parts of the metal surface are selectively etched. At present, commercial suppliers for chemical etching render the fabrication of metal structures attractive, by delivery at reasonable price and in relatively large quantities. Although details of the structured geometry vary for each material, a large variety of metals and stainless steel alloys can be utilized.

Despite these benefits in terms of process costs and flexibility, more severe restrictions for precision and shape variety are known for chemical etching as compared with other microstructuring techniques. Due to the isotropic nature of the etching process, only microstructures with small aspect ratios can be realized. In addition, structural dimensions are limited to a minimum of 300 µm.

Etching techniques were mainly employed to realize microchannel arrays for a number of gas phase reactors, including high temperature synthesis (see Figure 2-10) [72], catalyst screening [73] and periodically operated [49, 74] devices. In addition, stainless steel platelets were manufactured as parts of microextractors [8] and components of gas/liquid microreactors [75].

Interconnection can be realized simply by mechanical sealing applying pressure. In this case, small leakage streams will be tolerated as long as they do not strongly influence the reactor performance. More advanced sealing concepts are supplied by flat sealing techniques, e.g. based on graphite foils or special polymers. Irreversible sealing of metal foils can be achieved by laser welding and diffusion bonding. In the latter case, a special lamination technique was developed allowing a large number of thin microstructured metal laminates to be tightly joined [1, 2, 76].

At present, wet chemical etching can be considered to be the most advanced technology concerning mass fabrication of microreactors, especially with respect to costs and reliability. This is mainly due to an existing commercial delivery service in this field, supple-

25

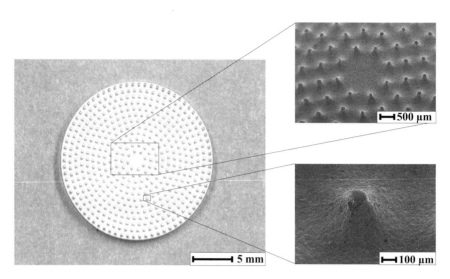

500 µm

5 mm

100 µm

Fig. 2-10. Photograph of a fin-type catalyst carrier element made of stainless steel by wet chemical etching [72].

mented by commercial suppliers of interconnection techniques for metal plates, e.g. referring to laser welding. The same does not hold either for punching or for embossing.

It is expected that the leading position of chemical etching will be put into perspective when further technological improvements can be achieved for the other techniques. Nevertheless, chemical etching currently provides a sound base to realize large-sized plates at affordable costs. This, in particular, is of interest to small- and medium-sized enterprises for microreaction technology that have a limited budget to undertake research efforts. In addition, the proof of mass fabrication capability already at the beginning of microreactor implementation paves the ground for the use of chemically etched production tools in small plants.

2.10 Electro Discharge Machining (EDM) of Conductive Materials

Another fabrication technology specifically suited for the processing of metals in industry is electro discharge machining (EDM) [30, 32, 77], which can be applied to all kinds of conductive materials. This fabrication method is well known from conventional machining in tool making [78]. EDM basically needs a voltage generator, an electrode and a workpiece. The technology is based on an erosion process, whereby conductive material is removed due to a high-energy discharge between an electrode and a workpiece which are surrounded by a dielectric fluid.

Extending EDM techniques to the micro-scale (µ-EDM) allows the generation of microstructures even with highly complex geometries independent of their mechanical properties [31, 79, 80]. Recently, μ-EDM proved its applicability as an appropriate tool for micro-component fabrication in microreaction technology [81].

μ-EDM technology is especially appropriate to realize microstructures in hard or chemically resistive materials. In this context, μ-EDM turns out to be a supplementing technique for the generation of metallic microstructures not realizable by other standard microstructuring techniques. Depending on the workpiece material employed during the EDM process, the resulting microstructures can be used as replication tools. For instance, these structures can be applied in embossing processes preferably at elevated temperatures.

Compared with conventional precision engineering, such as milling, turning or laser machining, EDM provides several advantages. Between electrode and workpiece no mechanical contact is made. The surfaces are not contaminated during manufacturing, hence eliminating cleaning procedures. Mechanical postprocessing is minimized since the edges of an eroded component can be prepared without any burr. Structural flexibility is higher due to the capability of three-dimensional microstructuring.

2.10.1 Wire-Cut Erosion and Die Sinking

Fabrication can be performed using sophisticated commercial wire-cut and die-sinking machines. In the first case, thin wires are used to cut platelets in order, preferably to generate electrodes for a die sinking process (see Figure 2-11) or to isolate single structures, e.g. from a wafer. In the die sinking process, a negatively shaped electrode, fabricated by LIGA, wire-cut erosion or any other technique, is generated in a workpiece by spark erosion.

As a special process variant, micro-die sinking with rotating disc electrodes of some ten of micrometers thickness, preferably of tungsten, was developed (see Figure 2-12) [82]. This process was also termed μ-EDM grinding. Thereby, microchannels of high aspect ratio providing large contact surfaces, e.g. for heat transfer, were machined.

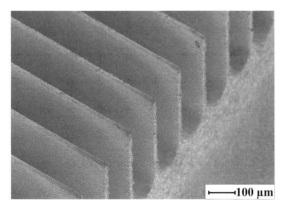

Fig. 2-11. Scanning electron micrograph of an electrode with thin comb-like structures made of tungsten by wire-cut erosion. This electrode is used to generate heat exchange microchannels by means of die sinking [49].

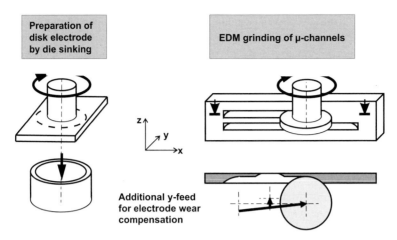

Fig. 2-12. Schematic of EDM grinding with rotating disc electrodes for generation of narrow and deep microchannels [82].

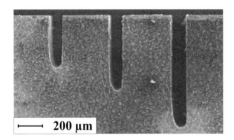

Fig. 2-13. EDM-ground microchannels of varying depth made of stainless steel. The thickness of the electrode was set to 25 μm [71].

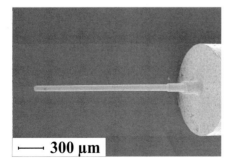

Fig. 2-14. Rod-like electrode for microdrilling EDM processes [72].

The central element of a liquid/liquid phase microreactor, a mixer/heat exchanger plate-let was realized by using the rotating disc technique. In Figure 2-13 a fluidic test structure is shown which served to demonstrate the feasibility of fabricating such high-aspect ratio microstructures.

2.10.2 μ-EDM Drilling

Another micro-EDM technique, called μ-EDM drilling, utilizes cylindrical electrodes of tungsten, hard alloys, tungsten/copper or brass with diameters between 10 μm and 100 μm, high speed electrode rotation (2000 rpm) and machine axis movement similar to conventional milling.

In addition, drilling techniques using small rod-like electrodes were developed, e.g. in order to manufacture inlet holes for a micromixing chamber or multichannel arrays in a catalyst structure (see Figure 2-14) [83, 84].

2.11 Laser Micromachining

Laser micromachining techniques were, in particular, useful for the generation of three-dimensional microstructures, since, different from conventional mask processes, they allow microstructuring in a vertical direction. Eliminating the need to use a mask, a fast transfer from CAD design to fabrication of a microstructure is achievable, referred to as fast prototyping. Consequently, laser processes principally allow one to adapt the design of a microstructure at short notice during the R&D phase.

Currently, this potential is widely unused, due to focusing of efforts to prove the feasibility of a concept for one process. However, the next stage of microreaction technology development will certainly require such a continuous design improvement based on simulation and experimental evaluation of the microstructures.

Laser micromachining was employed for the fabrication of membranes with isoporous holes designated for extraction purposes (see Figure 2-15) [76, 85]. In addition, ablation by

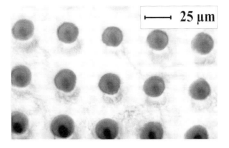

Fig. 2-15. Scanning electron micrograph of a membrane of a microextractor made of polyimide realized by excimer laser ablation [85].

using an excimer laser in the framework of a Laser-LIGA process [18] opened the gate to inexpensive realization of plastic micromixers [8, 18]. Laser irradiation was preferred over other kinds of high-energy exposure, e.g. using X-rays, since a three-dimensional profile of the mixer structure was needed in order to realize a certain new mixing principle.

A serial production of microcomponents by means of laser micromachining seems currently not to be easily attainable due to long processing times. However, this technique may become a versatile tool to optimize such structures prior to mass fabrication. Here, flexibility and fast transfer of concepts into products counterbalance the high specific costs of the fabrication process.

2.12 Interconnection Techniques

A number of interconnection techniques were applied in order to achieve fluid tightness in microreactors, depending on the construction material. For metals and stainless steels, diffusion bonding, electron beam welding, soldering, glueing and brazing were reported (see e.g. [68]). In addition to these irreversible interconnection techniques, reversible methods were applied such as mechanical sealing assisted by O-ring or graphite sealing (see e.g. [86]). Moreover, polymeric materials were joined by means of thermal bonding, glueing or laser welding (see e.g. [87, 88]).

2.12.1 Microlamination of Thin Metal Sheets

Since most studies concerning microreactors so far were focused on demonstrating processing feasibility, the development of interconnection techniques was only one working package among others to be carried out. Hence only a few publications refer exclusively to interconnection procedures, in particular regarding their impact on the shape of the fluidic structures and, hence, the fluid flow.

Microlamination Technique
A relatively detailed report was given by researchers of Oregon State University (U.S.A.) for the interconnection of thin metal sheets, termed microlamination [1, 74]. These sheets, referred to as shims or laminates as well, can principally be fabricated by any of the techniques described in Sections 4.3 – 4.10, preferably by ultraprecision machining, wet chemical etching, or laser cutting. The typically a few 100 μm thin sheets contain breakthrough microstructures, i.e. open channel systems, which need to be sealed by insertion of additional sheets, of the same or different material (see Figure 2-16).

Since the depth of the microchannels, which is equivalent to the sheet thickness, is much smaller than the channel width, the corresponding fluid structures are termed high-aspect-ratio microchannel (HARM) arrays. Please note that aspect ratio refers here only to a specific fluidic channel structure, characterized by large contact surfaces. However, it is not related to the microstructuring process itself, since the smallest dimension is

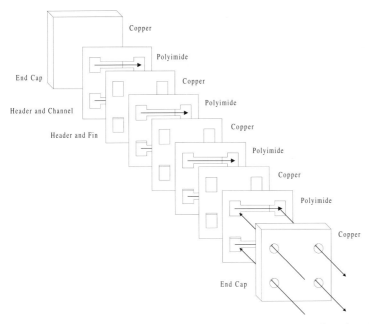

Fig. 2-16. Schematic of stacked sheets carrying fluidic structures assembled by using microlamination technique [2].

realized by means of microfabrication, and not the largest as, e.g., for LIGA microstructures.

For interconnection of HARM laminates, diffusion bonding, fusion welding, soldering, brazing, and adhesive bonding were employed. In the last case, microstructured polymer foils, e.g. polyimide, were inserted between the metallic laminates. For a microsystem (channels: 100 μm high, 3 mm wide, 10 mm long) consisting of polyimide and copper laminates, an adhesive bonding process was employed using a temperature of 265 °C and a pressure of 200 kPa for one minute. The interconnected foil stack is thereafter covered by two end caps which carry fluid in- and outlet connectors (see Figure 2-17).

Fluid Flow in Microlaminated Sheets
The channel width, amounting to about 3 mm, was reduced by 0.3 % after interconnection. As to be expected, the total height of the microdevice, which is proportional to the height of the microchannels, was reduced to a larger extent (about 4 %) due to compression of the polymeric foils. No clogging of the microchannels by insertion of polymeric foil material was observed.

The impact of the interconnection process on fluid flow in such assembled microdevices was investigated by a statistical analysis of the pressure drop carried out for four laminate stacks. For pressures ranging from 20 to 130 kPa, the average coefficient of variance of pressure drop was about 6 %. The corresponding flow rates were, on average, about 16 %

31

Fig. 2-17. Photographs of the HARM array fabricated by laser micromachining and adhesive bonding. Left: stack of laminates. Right: cross-section of laminate interface [2].

lower as predicted by theoretical calculations. Most likely, this is due to model simplifications, not taking into account entrance and exit flow effects in the manifold structures. An estimation of actual microchannel structure based on these deviations from ideal behavior suggests a maximum change of microchannel height of about 12 % or 12.6 μm absolute.

Microchannel Combustor/Reactor
One practical example of using the microlamination technique was given by the realization of a microchannel combustor/reactor for fuel processing with regard to proton exchange membrane (PEM) fuel cells, carried out by researchers at Pacific Northwest National Laboratory, Richland (U.S.A.) [1]. The device was divided into three parts, two of them consisting of microstructured laminates, to allow loading and replacement of metal foam catalyst supports.

Eight different shim patterns were needed to realize the complete fluidic system, comprising fluid lines for reactants and coolants. The complex system was composed of several laminate sets each consisting of two different shims arranged in alternating order. Thereby, different operational zones with effective internal heat transfer were defined. This set of shims comprises nearly 170 laminates, connected by diffusion bonding and was equipped with two endcaps for fluid connection. The combustor section consisted of two shim designs, also consisting of 170 laminates.

Gasoline Vaporizer
A gasoline vaporizer, another component of a fuel processor, was realized using two alternately stacked shim designs, forming two parallel series of microchannels, separated by 1.7 mm of solid metal [89]. The whole system consisted of 268 shims. Metal foam-supported combustion catalysts were inserted in open cavities of the shim stack. The hot gases resulting from combustion at the foam catalysts entered a series of microchannels, enabling efficient heat transfer to another series of microchannels carrying the gasoline to be evaporated. Thereby, a total gasoline flux of 265 scc/min could be vaporized.

32

Concluding Remarks

Apart from carrying out systematic studies to evaluate precise parameters for laminate interconnection, the work discussed above can be regarded as one of the first explicit reports on quality assurance with respect to microreaction technology. Assuming that mass production of microreactors is not far from its onset, this gives practical insight concerning the aptitude and limits of microstructuring and interconnection techniques as well as the extent of malfunction induced thereby. In this context, the research targets of this study are somewhat similar to the analysis of malfunctions caused by flow maldistribution [90]. Both studies showed, for realistic assumptions or even based on experimental data, that only relatively small deviations from ideality were observed, giving confidence that a reliable operation of many small microchannels ought to be possible.

2.13 Functional Coatings

The increased surface-to-volume ratio of microstructures, besides enhancing mass and heat transfer, may also induce some undesired effects, e.g. promoting corrosion or fouling. So far, no systematic study has been carried out in this field. However, a first feasibility study discussed the use of thin functional layers as coatings to increase corrosion stability and to reduce fouling of microchannel devices [91].

2.13.1 Functional Coatings for Corrosion Prevention

Researchers at the Forschungszentrum Karlsruhe (Germany) in co-operation with the KiKi Ingenieursgesellschaft, Karlsruhe (Germany) investigated the use of coatings to increase corrosion stability of metals and stainless steels as construction materials for microreactors [91]. They analyzed that in microchannels, due to high mass transfer, the corrosion potential generally is increased, thereby facilitating possible construction material damage. The researchers emphasized that functional coatings can only act to increase corrosion stability, not allowing full protection of an otherwise totally unstable material. Pinhole formation, inherent to all coating processes, renders a complete, long-term protection nearly impossible.

As one example, the formation of a passivating oxide layer for a special Fe-Al alloy, referred to as Fecralloy®, was investigated (see Figure 2-18). Mechanically microstructured samples were heated for 15 hours in air at about 1000 °C, yielding a 2–3 μm thin layer with aluminum oxide as the major fraction. This layer had a fine grainy surface texture. In order to demonstrate the feasibility of the concept, a further heating process over 15 hours was performed. This additional thermal treatment did not substantially increase the oxide layer thickness.

This concept is a first approach which has to be verified by further experiments. Tests under industrial standards, e.g. for periods as long as 1000 hours, and using real process gases are required.

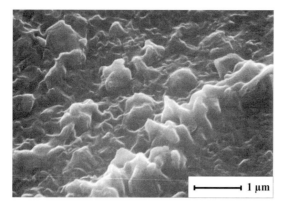

Fig. 2-18. Scanning electron micrograph of an aluminum oxide layer generated on the surface of an aluminum containing iron alloy [91].

2.13.2 Functional Coatings for Fouling Prevention

Fouling, in particular deposition of carbon layers from hydrocarbons, inside microchannels can pose severe problems by clogging of the fluid pathway, thereby increasing the pressure drop and potentially affecting the flow distribution. Moreover, the activity of catalyst surfaces may be notably reduced. Two principal approaches are likely to overcome fouling problems, namely generating extremely smooth surfaces or reducing the surface energy of construction materials. Following the latter route, a common procedure to realize very hydrophobic surfaces is given by fluorination or deposition of fluorinated layers on a substrate.

Fluorinated Antifouling Layers by ELBA and Sol–Gel Processes
Researchers at the Forschungszentrum Karlsruhe (Germany) employed in co-operation with the KiKi Ingenieursgesellschaft, Karlsruhe (Germany) a pulsed electron beam ablation (ELBA) and a sol–gel process for deposition of polytetrafluoroethylene-like layers on metals. Using the ELBA process, high kinetic energies of the material to be deposited can be utilized to enhance the mechanical adhesion of the layer. By this means, a uniform layer of 700 – 800 nm thickness was achieved on microstructured stainless steel foils. A fine structure of this fluorinated layer with characteristic dimensions in the range of several tens of nanometers was observed.

By a sol-gel method, a hydrolyzed mixture of two alkoxysilanes and a fluoroalkylsilane was employed for coating on the same types of substrates as used in the case of the ELBA process. After a dipping process in the sol solution, the final layer structure was achieved by heating at 150 °C for 10 hours. Good mechanical properties of such layers were evidenced by resistance against scratching and bending. The thickness of the fluorinated layer turned out to be slightly more than 2 μm. The fluorine content, being homogeneous through the layer, increased at the layer/metal interface, whereas carbon and oxygen had maximum concentrations at the air/layer interface. Similarly to the findings of the ELBA process, a fine structure was found.

Concluding Remarks

The results of employing the ELBA process clearly proved that a variation of surface energy is achievable by sol–gel coating as evidenced by a significant increase of the contact angle when immersed in water or methanol. However, more practically oriented tests are required to show the potential of such antifouling layers in detail. Only an experimental run of functionalized microstructures with a reaction of known fouling features, at best with industrial relevance, can bring deep insight into the real potential of this strategy.

2.14 References

[1] Matson, D. W., Martin, P. M., Steward, D. C., Tonkovich, A. L. Y., White, M., Zilka, J. L., Roberts, G. L.; *"Fabrication of microchannel chemical reactors using a metal lamination process"*, in Ehrfeld, W. (Ed.) *Microreaction Technology: 3rd International Conference on Microreaction Technology, Proceedings of IMRET 3*, pp. 62–71, Springer-Verlag, Berlin, (2000).

[2] Paul, B. K., Peterson, R. B., Wattanutchariya, W.; *"The effect of shape variation in microlamination on the performance of high-aspect ratio, metal microchannel arrays"*, in Ehrfeld, W. (Ed.) *Microreaction Technology: 3rd International Conference on Microreaction Technology, Proceedings of IMRET 3*, pp. 53–61, Springer-Verlag, Berlin, (2000).

[3] Gardeniers, J. G. E., Tjerkstra, R. W., van den Berg, A.; *"Fabrication and application of silicon-based microchannels"*, in Ehrfeld, W. (Ed.) *Microreaction Technology: 3rd International Conference on Microreaction Technology, Proceedings of IMRET 3*, pp. 36–44, Springer-Verlag, Berlin, (2000).

[4] Niggemann, M., Ehrfeld, W., Weber, L.; *"Micro molding of fluidic devices for biochemical applications"*, in Ehrfeld, W. (Ed.) *Microreaction Technology: 3rd International Conference on Microreaction Technology, Proceedings of IMRET 3*, pp. 113–123, Springer-Verlag, Berlin, (2000).

[5] Bremus, E., Gillner, A., Hellrung, D., Höcker, H., Legewie, F., Poprawe, R., Wehner, M., Wild, M.; *"Laserprocessing for manufacturing microfluidic devices"*, in Ehrfeld, W. (Ed.) *Microreaction Technology: 3rd International Conference on Microreaction Technology, Proceedings of IMRET 3*, pp. 80–89, Springer-Verlag, Berlin, (2000).

[6] Bengtsson, J., Wallman, L., Laurell, T.; *"High aspect ratio silicon micromachined heat exchanger"*, in Ehrfeld, W. (Ed.) *Microreaction Technology: 3rd International Conference on Microreaction Technology, Proceedings of IMRET 3*, pp. 573–577, Springer-Verlag, Berlin, (2000).

[7] Freitag, A., Dietrich, T. R.; *"Glass as a material for microreaction technology"*, in Proceedings of the "4th International Conference on Microreaction Technology, IMRET 4", pp. 48–54; 5–9 March, 2000; Atlanta, USA.

[8] Ehrfeld, W., Gärtner, C., Golbig, K., Hessel, V., Konrad, R., Löwe, H., Richter, T., Schulz, C.; *"Fabrication of components and systems for chemical and biological microreactors"*, in Ehrfeld, W. (Ed.) *Microreaction Technology, Proc. of the 1st Int. Conf. on Microreaction Technology*, pp. 72–90, Springer-Verlag, Berlin, (1997).

[9] Ehrfeld, W., Hessel, V., Möbius, H., Richter, T., Russow, K.; *"Potentials and realization of micro reactors"*, in Ehrfeld, W. (Ed.) *Microsystem Technology for Chemical and Biological Microreactors*, Vol. 132, pp. 1–28, Verlag Chemie, Weinheim, (1996).

[10] Wegeng, R. W., Call, C. J., Drost, M. K.; *"Chemical system miniaturization"*, in Proceedings of the "AIChE Spring National Meeting", 25–29 Febr., 1996; pp. 1–13; New Orleans,USA.

[11] Qin, D., Xia, Y., Rogers, J. A., Jackman, R. J., Zhao, X.-M., Whitesides, G. M.; *"Microfabrication, Microstructures and Microsystems"*, in Manz, A., Becker, H. (Eds.) *Topics in Current Chemistry*, Vol. 194, pp. 1–20, Springer Verlag, Heidelberg, (1998).

[12] Peterson, K. E.; in Proceedings of the "IEEE", pp. 420–457; 1982.

[13] James, J. B., Terry, S. C., Barth, P. W.; *"Silicon micromechanical devices"*, Sci. Am. **14**, p. 248 (1993).
[14] Rangelov, I. W.; *"Deep etching of silicon"*, Oficyna Wydawnicza Politechniki Wroclawskiej, Wroclaw (1996).
[15] Bhardwaj, J. K., Ashraf, H.; *"Advanced silicon etching using high density plasmas"*, SPIE **2639**, p. 224 (1995).
[16] Ehrfeld, W., Lehr, H.; *"Deep X-ray lithography for the production of three dimensional microstructures from metals, polymers and ceramics"*, Radiat. Phys. Chem. **45**, pp. 349–365, 3 (1995).
[17] Pfeufer, V.; *"Präzise Lichtblitze – Mikrobearbeitung mit dem Eximerlaser"*, F & M **104**, pp. 532–536, 7–8 (1996).
[18] Arnold, J., Dasbach, U., Ehrfeld, W., Hesch, K., Löwe, H.; *"Combination of excimer laser micromachining and replication processes suited for large scale production (Laser-LIGA)"*, Appl. Surf. Sci. **86**, p. 251 (1995).
[19] Bauer, H.-D., Weber, L., Ehrfeld, W.; *"Formeinsätze für die Massenfertigung von hochpräzisen Kunststoffteilen: Die LIGA-Technik"*, Werkzeug und Formenbau 4, pp. 22–26 (1994).
[20] Weber, L., Ehrfeld, W.; *"Micro-moulding-processes, moulds, applications"*, Kunststoffe plast europe **88**, pp. 10–63, 10 (1998).
[21] Friedrich, C., Warrington, R., Bacher, W., Bauer, W., Coane, P. J., Göttert, J., Hanemann, T., Haußelt, J., Heckele, M., Knitter, R., Mohr, J., Piotter, V., Ritzhaupt-Kreisel, H.-J., Ruprecht, R.; *"High Aspect Ratio Processing"*, in Rai-Choudhury, P. (Ed.) *Handbook of Microlithography, Micromachining and Microfabrication*, Vol. SPIE Monograph PM39/40; IEE Materials and Devices Series 12/12B, pp. 299–379, SPIE Optical Engineering Press, Washington, (1997).
[22] Dietrich, T. R., Ehrfeld, W., Lacher, M., Krämer, M., Speit, B.; *"Fabrication technologies for microsystems utilizing photoetchable glass"*, Micoelectron. Eng. **30**, pp. 497–504 (1996).
[23] Steckemetz, S., Hezel, R., Emmerthal, V.; *"Mikrostrukturierung mit der Draht-Läppmaschine"*, F & M **106**, pp. 146–149, 3 (1998).
[24] Kussul, E. M., Rachkovskij, D. A., Baidyk, T. N., A., Talayev, T. S.; *"Micromechanical engineering: a basis for the low-cost manufacturing of mechanical microdevices using microequipment"*, J. Micromech. Microeng. **6**, pp. 410–425 (1996).
[25] Westkämper, E., Hoffmeister, H.-W., Gäbler, J.; *"Spanende Mikrofertigung"*, F & M **104**, pp. 525–530, 7–8 (1996).
[26] Schubert, K., Bier, W., Linder, G., Seidel, D.; *"Profiled microdiamonds for producing microstructures"*, Ind. diamond rev. **50**, 5 (1990).
[27] König, W.; *Fertigungsverfahren Band 1: Drehen, Fräsen, Bohren*, VDI-Verlag, Düsseldorf (1990).
[28] Weck, M.; *Werkzeugmaschinen-Fertigungssysteme, Band 1: Maschinenarten, Bauformen und Anwendungsbereiche,* VDI-Verlag, Düsseldorf (1991).
[29] Striedieck, W.; *"Die Herstellung von präzisen Bauteilen durch Fotoätztechnik"*, Galvanotechnik.
[30] Dauw, D. E., van Coppenolle, B.; *"On the evolutions of EDM research – From fundamental to applied research"*, in van Griethuysen, J.-P. S., Kiritsus, D. (Eds.) *Proc. of International Symposium for Electromaching (ISEM XI)*, pp. 133–142, Presses Polytechniques et Universitaires Romandos, Lausanne, Switzerland, (1995).
[31] Ehrfeld, W., Lehr, H., Michel, F., Wolf, A.; *"Micro electro discharge machining as a technology in micromachining"*, in Proceedings of the "SPIE Symposium on Micromachining and Microfabrication", 14–15 Oct., 1996; pp. 332–337; Austin TX, USA.
[32] König, W.; *Fertigungsverfahren Band 3: Abtragen,* VDI-Verlag, Düsseldorf (1990).
[33] Zissi, S., Bertsch, A., Jézéquel, J. Y., Corbel, S., André, J. C.; Microsystem Technol. **2**, 97 (1996).
[34] Büttgenbach, S.; *Mikromechanik,* Teubner-Verlag, Stuttgart (1991).
[35] Heuberger, A.; *Mikromechanik,* Springer-Verlag, Berlin (1991).
[36] Menz, W., Mohr, J.; *Mikrosystemtechnik für Ingenieure,* 2nd ed; VCH, Weinheim (1997).
[37] Rai-Choudhury, P.; *"Handbook of Microlithography, Micromachining and Microfabrication"*, *SPIE Monograph PM39/40; IEE Materials and Devices Series 12/12B,* SPIE Optical Engineering Press, Washington, (1997).
[38] Shoji, S., Esashi, M.; *"Bonding and assembling methods for realising a µTAS"*, in Proceedings of the "Micro Total Analysis Systems, µTAS '94", 21–22 Nov., 1994; pp. 165–179; Twente, Netherlands.
[39] Lerou, J. J., Harold, M. P., Ryley, J., Ashmead, J., O'Brien, T. C., Johnson, M., Perrotto, J., Blaisdell, C. T., Rensi, T. A., Nyquist, J.; *"Microfabricated mini-chemical systems: technical feasibility"*, in Ehrfeld, W. (Ed.) *Microsystem Technology for Chemical and Biological Microreactors,* Vol. 132, pp. 51–69, Verlag Chemie, Weinheim, (1996).

[40] Jensen, K. F., Hsing, I.-M., Srinivasan, R., Schmidt, M. A., Harold, M. P., Lerou, J. J., Ryley, J. F.; *"Reaction engineering for microreactor systems"*, in Ehrfeld, W. (Ed.) *Microreaction Technology, Proceedings of the 1st International Conference on Microreaction Technology; IMRET 1*, pp. 2–9, Springer-Verlag, Berlin, (1997).

[41] Miyake, R., Lammerink, T. S. J., Elwenspoek, M., Fluitman, J. H. J.; *"Micromixer with fast diffusion"*, in Proceedings of the "IEEE-MEMS '93", Febr., 1993; pp. 248–253; Fort Lauderdale, USA.

[42] Branebjerg, J., Gravesen, P., Krog, J. P., Nielsen, C. R.; *"Fast mixing by lamination"*, in Proceedings of the "IEEE-MEMS '96", 12–15 Febr., 1996; pp. 441–446; San Diego, CA.

[43] Schwesinger, N., Frank, T., Wurmus, H.; *"A modular microfluid system with an integrated micromixer"*, J. Micromech. Microeng. **6,** pp. 99–102 (1996).

[44] Knight, J. B., Vishwanath, A., Brody, J. P., Austin, R. H.; *"Hydrodynamic focussing on a silicon chip: mixing nanoliters in microseconds"*, Phys. Rev. Lett. **80,** p. 3863, 17 (1998).

[45] Bökenkamp, D., Desai, A., Yang, X., Tai, Y.-C., Marzluff, E. M., Mayo, S. L.; *"Microfabricated silicon mixers for submillisecond quench flow analysis"*, Anal. Chem. **70,** pp. 232–236 (1998).

[46] Robins, I., Shaw, J., Miller, B., Turner, C., Harper, M.; *"Solute transfer by liquid/liquid exchange without mixing in micro-contactor devices"*, in Ehrfeld, W. (Ed.) *Microreaction Technology, Proceedings of the 1st International Conference on Microreaction Technology; IMRET 1*, pp. 35–46, Springer-Verlag, Berlin, (1997).

[47] Veser, G., Friedrich, G., Freygang, M., Zengerle, R.; *"A modular microreactor design for high-temperature catalytic oxidation reactions"*, in Ehrfeld, W. (Ed.) *Microreaction Technology: 3rd International Conference on Microreaction Technology, Proceedings of IMRET 3*, pp. 674–686, Springer-Verlag, Berlin, (2000).

[48] Franz, A. J., Quiram, D. J., Srinivasan, R., Hsing, I.-M., Firebaugh, S. L., Jensen, K. F., Schmidt, M. A.; *"New operating regimes and applications feasible with microreactors"*, in Ehrfeld, W., Rinard, I. H., Wegeng, R. S. (Eds.) *Process Miniaturization: 2nd International Conference on Microreaction Technology; Topical Conference Preprints*, pp. 33–38, AIChE, New Orleans, USA, (1998).

[49] Ehrfeld, W., Hessel, V., Löwe, H.; *"Extending the knowledge base in microfabrication towards themical engineering and fluid dynamic simulation"*, in Proceedings of the "4th International Conference on Microreaction Technology, IMRET 4", 5–9 March, 2000; Atlanta, USA.

[50] Gersten, K.; *Einführung in die Strömungsmechanik,* 6th ed; p. 98, Friedr. Vieweg & Sohn Verlagsgesellschaft mbH, Braunschweig (1991).

[51] McHardy, J., Ludwig, F.; *Electrochemistry of semiconductors and electronics,* Vol. 91, Noyes Publications, Rark Ridge (1994).

[52] Löwe, H., Ehrfeld, W., Hessel, V., Richter, T., Schiewe, J.; *"Micromixing technology"*, in Proceedings of the "4th International Conference on Microreaction Technology, IMRET 4", pp. 31–47; 5–9 March, 2000; Atlanta, USA.

[53] Ehrfeld, W., Bley, P., Götz, F., Hagmann, P., Maner, A., Mohr, J., Moser, H. O., Münchmeyer, D., Schelb, W., Schmidt, D., Becker, E. W.; *"Fabrication of microstructures using the LIGA process"*, in Proceedings of the "IEEE Micro Robots and Teleoperator Workshop", 9–11 Nov., 1987; p. 1; Hyannis, Cape Cod MA, USA.

[54] Ehrfeld, W., Abraham, M., Ehrfeld, U., Lacher, M., Lehr, H.; *"Materials for LIGA – products"*, in Proceedings of the "7th IEEE International Workshop on Micro Elektro Mechanical Systems", 25–28 Jan., 1994; Japan.

[55] Ehrfeld, W., Lehr, H.; *"Synchrotron radiation and LIGA-technique"*, Synchrotron Radiat. News **7,** pp. 9–13, 5 (1994).

[56] Löwe, H., Mensinger, H., Ehrfeld, W.; *"Galvanoformung in der LIGA-Technik"*, in Zielonka, A. (Ed.) *Jahrbuch Oberflächentechnik,* Vol. 50, pp. 77–95, Metall Verlag, Heidelberg, (1994).

[57] Niggemann, M., Ehrfeld, W., Weber, L., Günther, R., Sollböhmer, O.; *"Miniaturized plastic micro plates for application in HTS"*, Microsystem Technol. **6,** pp. 48–53, 2 (1999).

[58] Despa, M. S., Kelly, K. W., Collier, J. R.; *"Injection molding of polymeric LIGA HARMs"*, Microsystem Technologies **6,** pp. 60–66, 2 (1999).

[59] Ehrfeld, W., Hessel, V., Haverkamp, V.; *"Microreactors"*, *Ullmann's Encyclopedia of Industrial Chemistry,* Wiley-VCH, Weinheim, (1999).

[60] Kämper, K.-P., Ehrfeld, W., Döpper, J., Hessel, V., Lehr, H., Löwe, H., Richter, T.; *"Microfluidic components for chemical and biological microreactors"*, in Proceedings of the "MEMS-97", 26–30 Jan., 1997; p. 338; Nagoya, Japan.

[61] Knitter, R., Bauer, W., Fechler, C., Winter, V., Ritzhaupt-Kleissl, H.-J., Haußelt, J.; *"Ceramics in microreaction technology: materials and processing"*, in Ehrfeld, W., Rinard, I. H., Wegeng, R. S. (Eds.) *Process Miniaturization: 2nd International Conference on Microreaction Technology; Topical Conference Preprints*, pp. 164–168, AIChE, New Orleans, USA, (1998).

[62] Bauer, W., Knitter, R.; *"Formgebung keramischer Mikrokomponenten"*, Galvanotechnik **90**, 11 (1999) 3122–3130.

[63] Hessel, V., Ehrfeld, W., Freimuth, H., Löwe, H., Richter, T., Stadel, M., Weber, L., Wolf, A.; *"Fabrication and interconnection of ceramic microreaction systems for high temperature applications"*, in Ehrfeld, W. (Ed.) *Microreaction Technology, Proceedings of the 1st International Conference on Microreaction Technology; IMRET 1*, pp. 146–157, Springer-Verlag, Berlin, (1997).

[64] Winter, V., Knitter, R.; *"Al₂O₃/TiN as a material for microheaters"*, in Proceedings of the "Int. Conf. Micro Materials, MicroMat '97", 16–18 April, 1997; pp. 1015–1017; Berlin.

[65] Woolley, A. T., Mathies, R. A.; *"Ultra-high-speed DNA fragment separations using microfabricated capillary array electrophoresis chip"*, Proc. Natl. Acad. Sci. USA **91**, pp. 11348–11352 (1994).

[66] Hessel, V., Ehrfeld, W., Möbius, H., Richter, T., Russow, K.; *"Potentials and realization of micro reactors"*, in Proceedings of the "International Symposium on Microsystems, Intelligent Materials and Robots", 27–29 Sept., 1995; pp. 45–48; Sendai, Japan.

[67] Freitag, A., Dietrich, T. R.; *"Glass as a material for microreaction technology"*, in Proceedings of the "4th International Conference on Microcreation Technology, IMRET 4", 5–9 March, 2000; pp. 48–54; Atlanta, USA.

[68] Schubert, K.; *"Entwicklung von Mikrostrukturapparaten für Anwendungen in der chemischen und thermischen Verfahrenstechnik"*, KfK Ber. **6080**, pp. 53–60 (1998).

[69] Ehrfeld, W., Hessel, V., Kiesewalter, S., Löwe, H., Richter, T., Schiewe, J.; *"Implementation of microreaction technology in process engineering and biochemistry"*, in Ehrfeld, W. (Ed.) *Microreaction Technology: 3rd International Conference on Microreaction Technology, Proceedings of IMRET 3*, pp. 14–35, Springer-Verlag, Berlin, (2000).

[70] Knitter, R., Günther, E., Odemer, C., Maciejewski, U.; Microsystem. Technol. **2**, pp. 135–138, 3 (1996).

[71] Richter, T., Ehrfeld, W., Wolf, A., Gruber, H. P., Wörz, O.; *"Fabrication of microreactor components by electro discharge machining"*, in Ehrfeld, W. (Ed.) *Microreaction Technology, Proc. of the 1st International Conference on Microreaction Technology*, pp. 158–168, Springer-Verlag, Berlin, (1997).

[72] Results of IMM, unpublished

[73] Zech, T., Hönicke, D., Lohf, A., Golbig, K., Richter, T.; *"Simutaneous screening of catalysts in microchannels: methodology and experimental setup"*, in Ehrfeld, W. (Ed.) *Microreaction Technology: 3rd International Conference on Microreaction Technology, Proceedings of IMRET 3*, pp. 260–266, Springer-Verlag, Berlin, (2000).

[74] Liauw, M., Baerns, M., Broucek, R., Buyevskaya, O. V., Commenge, J.-M., Corriou, J.-P., Falk, L., Gebauer, K., Hefter, H. J., Langer, O.-U., Löwe, H., Matlosz, M., Renken, A., Rouge, A., Schenk, R., Steinfeld, N., Walter, S.; *"Periodic operation in microchannel reactors"*, in Ehrfeld, W. (Ed.) *Microreaction Technology: 3rd International Conference on Microreaction Technology, Proceedings of IMRET 3*, pp. 224–234, Springer-Verlag, Berlin, (2000).

[75] Hessel, V., Ehrfeld, W., Golbig, K., Haverkamp, V., Löwe, H., Storz, M., Wille, C., Guber, A., Jähnisch, K., Baerns, M.; *"Gas/liquid microreactors for direct fluorination of aromatic compounds using elemental fluorine"*, in Ehrfeld, W. (Ed.) *Microreaction Technology: 3rd International Conference on Microreaction Technology, Proceedings of IMRET 3*, pp. 526–540, Springer-Verlag, Berlin, (2000).

[76] Martin, P. M., Matson, D. W., Bennett, W. B.; *"Microfabrication methods for microchannel reactors and separation systems"*, in Ehrfeld, W., Rinard, I. H., Wegeng, R. S. (Eds.) *Process Miniaturization: 2nd International Conference on Microreaction Technology; Topical Conference Preprints*, pp. 75–80, AIChE, New Orleans, USA, (1998).

[77] Christmann, H.; *"Feinsterosion in der Praxis"*, in Proceedings of the "Microengineering '96, Spezialthema: Mikrobearbeitung von Metallen", 11–13 Sept., 1996; p. 33; Stuttgart, Germany.

[78] Locher, A., Gruber, H. P.; *"µ-EDM – Technik oder Kunst?"*, in Proceedings of the "Microengineering '96, Spezialthema: Mikrobearbeitung von Metallen", 11–13 Sept., 1996; pp. 27–32; Stuttgart, Germany.

[79] Wolf, A., Berg, U., Ehrfeld, W., Lehr, H., Michel, F., Zimmerschitt, N., Gruber, H. P.; *"Feine Sache: Die Mikrofunkenerosion ermöglicht beim Präzisionsspritzguß sehr filigrane Strukturen"*, Der Zuliefermarkt 4, pp. 48–52, (1997).

[80] Wolf, A., Lehr, H., Nienhaus, M., Michel, F., Gruber, H. P., Ehrfeld, W.; *"Combining LIGA in EDM for the generation of complex microstructures in hard materials"*, in Proceedings of the "Progress in Precision Engineering and Nanotechnology, 9-IPES/UME4 Conf. '97", 26–30 May, 1997; pp. 657–660; Braunschweig, Germany.

[81] Jäckel, K., P., Wörz, O.; *"Winzlinge mit großer Zukunft-Mikroreaktoren für die Chemie"*, Chemie-Technik **26,** pp. 130–134, 1 (1997).

[82] Wolf, A., Ehrfeld, W., Lehr, H., Michel, F., Richter, T., Gruber, H., Wörz, O.; *"Mikroreaktorfertigung mittels Funkenerosion"*, F & M, Feinwerktechnik, Mikrotechnik, Meßtechnik 6, pp. 436–439 (1997).

[83] Hessel, V., Ehrfeld, W., Golbig, K., Hofmann, C., Jungwirth, S., Löwe, H., Richter, T., Storz, M., Wolf, A., Wörz, O., Breysse, J.; *"High temperature HCN generation in an integrated Microreaction system"*, in Ehrfeld, W. (Ed.) *Microreaction Technology: 3rd International Conference on Microreaction Technology, Proceedings of IMRET 3,* pp. 151–164, Springer-Verlag, Berlin, (2000).

[84] Löwe, H., Ehrfeld, W.; *"State of the art in microreaction technology: concepts, manufacturing and applications"*, Electrochim. Acta **44,** pp. 3679–3689 (1999).

[85] Tegrotenhuis, W. E., Cameron, R. J., Butcher, M. G., Martin, P. M., Wegeng, R. S.; *"Microchannel devices for efficient contacting of liquids in solvent extraction"*, in Ehrfeld, W., Rinard, I. H., Wegeng, R. S. (Eds.) *Process Miniaturization: 2nd International Conference on Microreaction Technology; Topical Conference Preprints,* pp. 329–334, AIChE, New Orleans, USA, (1998).

[86] Richter, T., Ehrfeld, W., Gebauer, K., Golbig, K., Hessel, V., Löwe, H., Wolf, A.; *"Metallic microreactors: components and integrated systems"*, in Ehrfeld, W., Rinard, I. H., Wegeng, R. S. (Eds.) *Process Miniaturization: 2nd International Conference on Microreaction Technology, IMRET 2; Topical Conference Preprints,* pp. 146–151, AIChE, New Orleans, USA, (1998).

[87] Konrad, R., Ehrfeld, W., Hartmann, H.-J., Jacob, P., Neumann, M., Pommersheim, R., Sommer, I., Wolfrum, J.; *"Disposable electrophoresis chip for high throughput analysis of biomolecules"*, in Ehrfeld, W. (Ed.) *Microreaction Technology: 3rd International Conference on Microreaction Technology, Proceedings of IMRET 3,* pp. 420–429; Springer-Verlag, Berlin, (2000).

[88] McReynolds, R. J.; *"Method of Manufacturing Microfluidic Devices"*, US 5882465, (18.06.1997); Caliper techologies Corp.

[89] Tonkovich, A. L., Fitzgerald, S. P., Zilka, J. L., LaMont, M. J., Wang, Y., VanderWiel, D. P., Wegeng, R.; *"Microchannel chemical reactor for fuel processing applications. – II. Compact fuel vaporization"*, in Ehrfeld, W. (Ed.) *Microreaction Technology: 3rd International Conference on Microreaction Technology, Proceedings of IMRET 3,* pp. 364–371, Springer-Verlag, Berlin, (2000).

[90] Walter, S., Frischmann, G., Broucek, R., Bergfeld, M., Liauw, M.; *"Fluiddynamische Aspekte in Mikroreaktoren"*, Chem. Ing. Tech. **71,** pp. 447–455, 5 (1999).

[91] Fichtner, M., Benzinger, W., Hass-Santo, K., Wunsch, R., Schubert, K.; *"Functional coatings for microstructure reactors and heat exchangers"*, in Ehrfeld, W. (Ed.) *Microreaction Technology: 3rd International Conference on Microreaction Technology, Proceedings of IMRET 3,* pp. 90–101, Springer-Verlag, Berlin, (2000).

3 Micromixers

3.1 Mixing Principles and Classes of Macroscopic Mixing Equipment

Mixing is a physical process with the goal of achieving a uniform distribution of different components in a mixture, usually within a short period of time. This definition includes the integration of two or more fluids into one phase, most often accompanied by volume contraction or expansion, or the interdispersion of solids. Concerning fluids, the contacting of miscible and non-miscible media has to be taken into account. In the latter case gas/liquid dispersions and emulsions are obtained.

Heat Release or Consumption During Mixing
For a number of reasons heat may be released or consumed during mixing processes, e.g. due to mixing thermodynamics, energy transfer from the mixing equipment or by chemical reaction. Positive or negative mixing enthalpies are a consequence of changes of the molecular environment, e.g. a variation of van der Waals, dipole or ionic forces, and result either in an endothermic or exothermic behavior. Another origin of heat transfer is given by the inevitable need for actively assisting mixing processes, i.e. the interaction of the fluids or solids with the mixing equipment, e.g. the stirrer or pump. The mechanical, pressure, vibration or flow energy introduced is partially converted into heat. This influence is, at least, not negligible for some mixing devices which need large quantities of energy to achieve mixing, e.g. in the range of several tens to hundreds of kW per 100 l of mixture. Furthermore, if mixing is coupled with chemical reaction, huge heat releases can occur, in particular for fast reactions having large reaction enthalpies.

Molecular Diffusion
Among the mechanisms known for mixing of liquid or gaseous phases, molecular diffusion is the final step in all mixing processes. Diffusional transport obeys Fick's law which correlates to the change of concentration with time, and hence to the product of the diffusion coefficient and concentration gradient [1]. By rearrangement of this relationship, the following equation results:

$$t \sim d_l^2 / D \qquad (3.1)$$

D: diffusion coefficient
d_l: lamella width
t: mixing time

Mixing therefore depends on the diffusion constant D and the diffusional path d. Apart from high-molecular mass species such as polymers, typical diffusion constants of liquids or dissolved solids do not differ much. For instance, the contacting of two 500 μm thick aqueous layers results in mixing times of several minutes [2]. Thus, diffusion generally is, without any additional support, a rather slow process and has to be actively assisted by division of the mixing volume into small compartments or thin lamellae, i.e. decreasing the diffusional path d. This rule even applies to gases although their diffusion constants are about three orders of magnitude higher compared with liquids.

Turbulent and Laminar Mixing
Two concepts of volume division of liquids and gases have been followed: creation of turbulent flow and laminar mixing [3, p.4,4, pp.10–12]. In a turbulent regime the fluid entity constantly is subdivided into thinner and thinner layers by an induced circular motion of fluid compartments, so-called eddies, and subsequent breaking into fragments (see Figure 3-1).

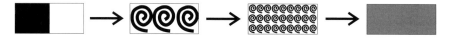

Fig. 3-1. Schematic of the creation of eddies by stirring (turbulent mixing).

In a laminar regime, a similar breaking of fluid compartments cannot occur due to the high viscous forces. Instead, the fluid entity has to be continuously split and recombined, forming regularly-sized fluid embodiments (see Figure 3-2). To give a vivid example, this process is similar to the cutting of slices from two different solid blocks, and subsequent alternate stacking of the slices to give a joint assembly. Multiple repetition of this procedure finally leads to a dispersed solid body.

Fig. 3-2. Schematic of the splitting by continuous geometric separation and reunion of fluid embodiments (laminar mixing).

Mixing of solids is considerably different from that of fluids since two entities do not have to be dispersed, but rather an interpenetration of hard, indivisible pieces to some kind of conglomerate has to be achieved. For this purpose different mechanisms, termed distributive and dispersive, have been described [3, p.4].

Mixing Equipment
A huge number of diverse types of mixing apparatus for fluids and solids use the two basic mixing concepts mentioned above, which can be categorized in several classes. One suitable classification was introduced by Wilke, Buhse and Groß [3, p.3] which relates the mixer classes to the type of energy input used to stimulate mixing:

- Mechanically operating mixers
- Streaming mixers
- Pneumatic mixers
- Vibrating mixers
- Special-type mixers

Mechanically operating mixers certainly form the most prominent class of mixers. Various process variants of mechanical operation have been developed such as compulsion, homogenization, shaking and free-falling concepts [3, p.21]. While these concepts are mainly applied to solid mixing, mechanical stirring is a commonly known variant for mixing of low-viscous liquids by induced turbulent flow.

Various impellers were developed to improve the flow pattern of a liquid in a vessel and to be specially adapted to a number of mixing problems [4, pp.78–91,5]. Paddle or propeller impellers are major components of low-viscosity mixing equipment utilizing turbulence, while helical impellers have been used for laminar mixing, e.g. mixing of viscous liquids, pastes and creams.

Another method applied for turbulence generation is based on guided streaming of flows, preferably leading to direct collision. One class of this so-called *streaming mixers*, uses free turbulence, achieved by generation of a high flow velocity in mixing elements having a suitable shape [3, pp.145–177]. Other types of *streaming mixers* are based on the principle of splitting and recombination of the streams to be mixed, by either layer or eddy formation dependent on the Reynolds number. Mixing nozzles are examples of *streaming mixers* using free turbulence, while mixing pumps and static mixers [6] perform continuous layer splitting and recombination.

While for mixing of liquids and gases mechanically operating and streaming mixers are widely used, for solids still other types of apparatus have been developed. Pneumatic mixers are designed to fluidize solid bulk materials [3, pp.133–144]. Vibrating mixers, e.g. vortex mixers, transfer vibrations from an external source into a mixture, which is often accompanied by bubble generation of air introduced e.g. via a tube [3, pp.179–186]. Silo mixers remove material at various positions within a solid and rejoin the parts taken [3, pp.187–198]. Finally, a number of mixers do not fit into the categories mentioned above and, hence, have to be characterized, according to this definition, as special-type mixers.

3.2 Mixing Principles and Classes of Miniaturized Mixers

Since micromixers have in almost all cases a laminar flow regime due to the small channel dimensions, the vast majority are based on diffusional mixing without any assistance of turbulence. This process is in most cases performed between thin fluid layers, although other types of fluid compartments have been applied as well (listed below). The formation of thin layers is achieved simply by division of a main stream into many small substreams or reduction of the channel width along the flow axis for one channel. Thereby, large contact surfaces and small diffusional paths are generated.

Although diffusion is the dominant mechanism and already quite efficient on a microscale, a number of micromixers make use of assisting streaming phenomena to improve mixing. These phenomena are similar to those mentioned in Section 3.1. Multiple redivision of a layered system can be achieved via splitting and recombination mechanisms of fluid segments. Bending, drilling and turning concepts have been applied as well [7–9]. Thereby, the diffusional path is strongly reduced. To give a concrete example, diffusion of a small organic molecule in an aqueous system needs about 5 s for a path of 100 µm, while only 50 ms are required to pass a 10 µm thin lamellae [2].

Moreover, a number of further mechanisms assisting diffusion may principally be applied, e.g. the use of mechanical, thermal, vibrational and electrical energy [3, p.3], and some of them have been reported already. For instance, ultrasonic wave generators can be used to generate vibrations as well as actuators for a mechanical displacement of fluid layers.

Although being an exception from the majority of devices operating in the laminar regime, the use of turbulent mixing has been described in a few cases. For instance, stirring in microsystems by the use of moving magnetic beads was reported [10]. In addition, at very high volume flows the occurrence of turbulent conditions was claimed for miniaturized mixers with dimensions which are at the borderline to conventional tubes, i.e. diameters of several hundred micrometers [11].

On the basis of these mixing concepts, the following generic flow arrangements can be listed, each defining a separate type of miniaturized contacting device (see Figure 3-3).

1) Contacting of two substreams, e.g. in a mixing tee configuration
Although simple in design, such devices are useful mixers, if the dimensions are sufficiently small, and, hence, have been applied in a number of cases. In the case of larger channels and flows of high velocities, mixing assisted by turbulence has been reported [11].

2) Collision of two substreams of high energy and generation of a large contact surface due to spraying/atomizing
One example of this flow configuration refers to a mixing device consisting of a spherical mixing chamber the center of which forms the focal point of the directions of three triangular arranged small openings, manufactured by means of ultraprecision machining [13, 14]. Through these openings three liquid streams enter at supersonic velocity and jointly collide in the center. One field of application refers to crystallization to small particles using the extremely high energy released. The latter is so high that any contact of one stream with a wall of the mixing chamber, as a consequence of dejustification, would remove construction material. In order to avoid such damage, large ceramic spheres are inserted in the mixing chamber.

3) Injection of many small substreams of one component into a main stream of another component
Sieve- or slit-like mixing devices belong to this category. Actually, one of the first micromixers described in the literature was based on this principle [15]. A detailed description of this device is given in Section 3.5.

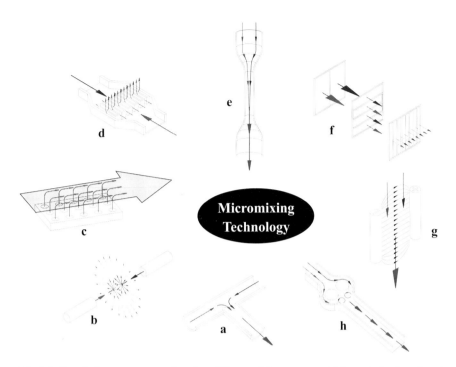

Fig. 3-3. Generic flow arrangements based on different mixing concepts yielding a design base for micromixer development [12]. **a**: Contacting of two substreams, **b**: Collision of two high energy substreams, **c**: Injection of many small substreams of one component in a main stream of another component, **d**: Injection of many substreams of two components, **e**: Decrease of diffusion path perpendicular to the flow direction by increase of flow velocity, **f**: Manifold splitting and recombination, **g**: Externally forced mass transport, e.g. by stirring, ultrasonic wave, or electrical energy, **h**: Periodic injection of small fluid segments.

4) Injection of many substreams of two components

This approach has been used by a number of authors and has proven to result in good mixing in a number of cases [2, 16, 17]. Therefore, a profound knowledge base has been gained for mixers of this type and some devices are commercially available. In addition, multiphase mixing seems to be of increasing importance for such mixers [18].

5) Decrease of diffusion path perpendicular to the flow direction by increase of flow velocity

One mixing device based on this principle has been reported and tested for applications in biochemistry [19].

6) Manifold splitting and recombination of a stream consisting of two fluid lamellae of two components

Actually, this is a well-known principle applied in many macroscopic static mixers [3, pp.161–174,6]. A number of authors have developed different strategies to realize the process sequence mentioned above. Thus, a variety of mixers meanwhile has been realized and tested. In addition, various efforts for simulating the mixing process in these devices have been made [9].

However, apart from simple test reactions or applications in biochemisty, hardly any experiments have been directed towards chemical engineering. Furthermore, the multiphase behavior of the devices has rarely been described.

7) Externally forced mass transport, e.g. by stirring, ultrasonic wave, electrical and thermal energy

One example of stirred mixing in microstructures has been given [10]. There are hardly any reports about using ultrasonic waves or of electrical and thermal energy.

8) Periodic injection of small fluid segments

Periodic injection can be performed by several means. Possible routes are alternating working microvalves or injection through two hole arrays similar to ink jet printers which differ in row position. Furthermore, fluid volumina may be consecutively arranged in the gaps of the cogs of a gear pump or by membrane motions of two membrane pumps which differ by half a period [20].

The characterization of mixing in devices of these types has, to our best knowledge, not been reported widely. However, the fabrication of a number of such devices, e.g. micropumps, valves and hole arrays, has been practised for about ten years. Thus, it seems to be reasonable and worthwhile to modify these devices according to the needs of mixing processes.

9) Mixed contacting types mentioned in 1) to 8)

Although, at present, detailed characterization and comparison of mixing devices based on a single mechanism are needed, it is very likely that full performance may only be achieved by the combined action of several concepts in one complex microstructured device.

3.3 Potential of Miniaturized Mixers

Apart from the general goal of mixing, to achieve a uniform distribution of components within a certain time period (see Section 3.1), micromixers offer features which cannot be achieved by macroscopic devices.

Ultrafast Mixing on Macro-Scale

First, some specialized micromixers, capable of generating thin layers on the nanometer scale, allow ultrafast mixing of low-viscous solutions down to a few microseconds [19], which is far beyond the limit of standard macroscopic equipment. It should be mentioned that these devices, owing to their mixing principle, are limited to certain applications (see detailed discussion in Section 3.8.1).

Although generally slower than these special mixers, the overwhelming majority of micromixers are still capable of performing mixing, at least in the range below one second

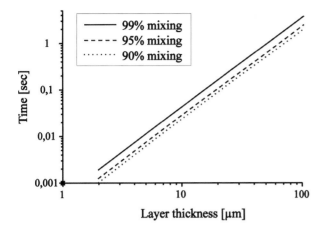

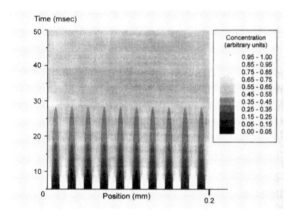

Fig. 3-4. Top: calculated mixing times versus layer thickness of a micromixer, which equals to channel width. Bottom: visualization of the dynamic development of a mixing process [2].

down to a few milliseconds. These values are estimated values from a published correlation [2], based on Fick's law, revealing the dependence of the mixing time t on the width of the lamellae d_l (see Figure 3-4 and also Equation 3.1).

Since the lamellae width is not known in most mixing applications, its value is usually assumed, in a first approach, to be equivalent to the sum of the width of typical microstructures, e.g. channels or nozzles, and wall width, the latter separating the microstructures. Typical length scales for microstructures range from a few to hundreds of micrometers, while walls are made as thick as necessary, usually not extending a few tens of micrometers. Whenever researchers were able to verify mixing times experimentally, the data determined were close to the calculated values or, at least, of the same order [9]. This underlines that mixing by molecular diffusion is the dominat mechanism when operating in a laminar regime.

Homogeneous Mixing on Micro-Scale

Secondly, since micromixers allow remarkably precise control of the fluid layer thickness, they permit a defined setting of the diffusional path. This absolutely uniform geometrical situation at the start of mixing leads to a very narrow distribution of local mixing times of the different fluid compartments within the mixing volume. This situation is totally different from a stirred vessel having zones of high mechanical agitation close to the impeller and, hence, small eddies compared to "quieter zones" outside with larger eddies.

The control over layer thickness will be used to generate small lamellae in the majority of applications. However, if a slow mixing process is demanded, e.g. for highly exothermic reactions with large heat release, the formation of layers of medium thickness may result in a better overall process performance. For aqueous lamellae of 100 µm thickness containing small solute molecules, for instance, mixing times of the order of seconds are achieved.

Certainly, slow mixing can be achieved in macroscopic equipment as well. However, "slow" refers here to the average mixing time of the whole apparatus, still consisting of fast and very slow local time scales. In the micromixer the average mixing time equals the local values since a homogeneous fluid geometry, i.e. equally partitioned layers, is provided all over the device at the start of mixing.

Integration of Mixing Function in Complex System

A third argument for the use of micromixers is that, due to system miniaturization and integration, a close connection to other miniaturized components, e.g. micro heat exchangers, can be obtained. If, for one of the reasons listed in Section 3.1, heat is released during mixing, this often has a negative impact on the quality of the mixing products, e.g. by inducing side or follow-up reactions as well as thermal degradation of the mixture. The ability to transfer heat at the place of generation, quasi in-situ and in-line, yields a high performance of thermal control which is not achievable in any standard equipment.

Gas/Liquid Dispersions and Emulsions

Finally, analyzing the potential of micromixers, the fast growing field opened by contacting of non-miscible media has to be taken into account. First stimulating results prove the generation of suprisingly uniform dispersions and emulsions in micromixers [21–23]. Very small bubbles and droplets were observed corresponding to large specific interfaces up to 50,000 m²/m³. The few applications reported range from emulsion and paste formation to multiphase and phase transfer reactions.

In contrast to homogeneous mixing, no detailed theoretical explanation has been introduced for the description of multiphase formation in micromixers to date. However, even for large-scale devices, simulations of these phenomena are complex and sometimes not possible at all. Therefore, the interpretation of first results has to be made cautiously and much more data are needed for a knowledge base for further micromixer design.

3.4 Contacting of Two Substreams, e.g. in a Mixing Tee Configuration

3.4.1 Mixing Tee-Type Configuration

Mixing in Straight and Meandering Channels
A fundamental study investigating the influence of channel geometry on the mixing behavior of bilayered streams was performed by researchers at the Mikroelektronik Centret in Copenhagen (Denmark) [23]. In a first experiment, the shape of the channels was varied. The mixing performance of a straight channel was compared with a meandering one (see Figure 3-5). In a second experiment, the hydraulic diameter of the channels was reduced. Both variations influence the Reynolds number and, hence, the flow regime of the liquids. The transition from a laminar to a turbulent regime has a direct impact on the mixing performance since molecular diffusion is assisted by induced motion and fragmentation of fluid compartments.

Microfabrication
The mixer devices were fabricated in a transparent acrylic block using conventional precision machining. Two sets of straight and meandering channels with different hydraulic diameter were fabricated. For the first set, having the larger diameter, the dimensions of both channels were equal. A channel length of 100 mm was chosen and the depth and width were fixed to 300 μm and 600 μm. The second set consisted of much smaller channels, 280 μm wide and 25 μm deep. The straight channel had a length of 5 mm, while the meandering channel was 25 mm long and consisted of ten curves.

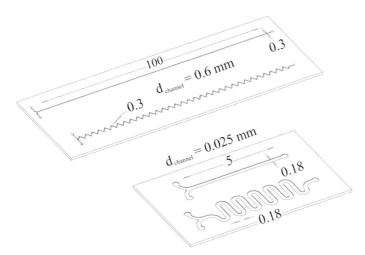

Fig. 3-5. Schematic of two sets of mixing tee-type structures, each consisting of straight and meandering channels [23].

Mixing in Large-Sized Straight and Meandering Channels
The evaluation of the mixing process was carried out by monitoring the flow of colored liquids. Mixing of a solution of the slightly yellow pH indicator bromothymol blue with sodium hydroxide resulted in a dark blue solution.

In the first experiment, only incomplete mixing was observed for the straight channel. A color change could be noticed only directly at the fluid interface. These results are in accordance with the long time scale of the diffusion processes. Diffusion is expected to be the only mechanism contributing to mixing for the straight channel because a laminar regime results for all flow rates. In contrast, for the meandering channel a transition to a turbulent regime occurs at high flow rates, restricting the laminar flow to low flow rates. For instance, at a flow rate of 4 ml/min mixing was completed already after one third of the channel length.

Mixing in Small-Sized Straight and Meandering Channels
In the second experiment, mixing was performed in the smaller channels. As to be expected, the flow regimes were generally restricted to the laminar regime, independent of the shape of the channels. Thus, mixing was observed only at the interface between the two liquids, even when the flow passed very sharp corners. The colored mixing zone around the interface broadened down the channel, i.e. demonstrating the increasing contribution of molecular diffusion for longer residence times. If the latter was further increased by lowering the flow rate, an even broader mixing zone resulted.

Based on calculations of the Reynolds number and the experimental results, the authors demonstrated that non-laminar flow can occur even at Reynolds numbers as low as 15, as far as the flow in short channels or orifices is concerned. Here, transitional Reynolds numbers have to be applied rather than the constant value of 2300 known for established flow profiles in long channels. A calculation of the transitional number for the large meandering channel shows that turbulent mechanisms were likely to contribute to mixing. The Reynolds numbers of the flow regimes of the three other microchannels were far from the transition value.

3.4.2 Double Mixing Tee-Type Configuration

Quench-Flow Analysis
Researchers at the Caltech Micromachining Laboratory and the Howard Hughes Medical Institute in Pasadena (U.S.A.) realized a mixing device as a measuring tool for studying fast consecutive reactions and processes. This study is, in particular, important in the field of quench-flow and stop-flow analysis, e.g. to monitor protein folding, where a time resolution of less than one millisecond is required [11]. Therefore, a sudden start and stop of processes and reactions by intimate mixing has to be ensured. Moreover, the volume of the mixing chamber, which usually is quite large for macroscopic mixing devices, has to be strongly reduced in order to minimize dead volumes and, hence, dead times.

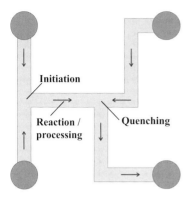

Fig. 3-6. Schematic representation of the double mixing tee-type arrangement of a micromixer for sub-millisecond quench-flow analysis [11].

Double Mixing Tee Microdevice

The mixing device comprises a microchannel system which resembles two mixing tees connected in series (see Figure 3-6). The outlet of the first mixing tee forms one inlet of the second mixing tee, and the other inlet is fed by an external source. In the second mixing tee another fluid is introduced and mixed with the mixture of the first two fluids.

Basically, this geometry provides a residence time channel of defined length separated by two mixing tees. In the idealized case of indefinitely short mixing, this configuration would enable one immediately to start a reaction and to control reaction time by variation of flow rate or length of flow passage before stopping the reaction by a second mixing process. Practically, a major part of the volume of the residence time channel will, even if good mixing is ensured, be demanded by mixing issues rather than by needs of reaction (as proven by the results below).

Operation at high volume flows is expected to guarantee a reasonably good mixing due to a favorable use of turbulences, which are induced by direct collision of two streams. Thus, for the purpose claimed, the study of ultrafast processes, the double-tee mixer is perfectly suited. However, at lower flows the performance should significantly decrease, thereby restricting the mixer to certain applications. In contrast, laminar-operating micromixers, using molecular diffusion, should ideally reveal a constant mixing quality over a broad range of volume flows. Indeed, first results underline this expectation [16].

Microfabrication

The microchannel system was fabricated by standard silicon micromachining via etching of a 10 x 10 mm^2 silicon chip with potassium hydroxide using thermal oxide as an etch mask. The double mixing tee configuration consists of six microchannels of 400 µm width and about 2 mm length. For fluid connection, an outlet hole of 750 µm diameter, was drilled into the silicon chip. The chip was anodically bonded to a glass slide with three inlet holes,

51

clamped in a holder and, thereby, connected to a commercially available quench-flow instrument.

Mixing Characterization/Basic Hydrolysis of Phenyl Chloroacetate
In feasibility experiments the time scale of mixing was determined and compared with that of the reactions to be investigated. The first and second mixing processes were followed by flow visualization using two-fold color changes of a pH indicator containing solution. In a first step, mixing with an acid resulted in a color change. Second, the original color was re-established by subsequent addition of a base.

In order to vary residence times and mixing performance, flow rates in the range of 1.8 to 9.0 l/h were applied, corresponding to Reynolds numbers of 1000 up to 8000. A fully developed color profile, proving high mixing quality, along the channel axis of the mixer was only observed for flow rates as high as 2.7 l/h. For smaller rates, e.g. 0.7 l/h, only incomplete mixing resulted.

On the basis of these results, a flow range of 1.8 to 9.0 l/h was chosen for the investigation of the basic hydrolysis of phenyl chloroacetate as a test reaction for quench-flow analysis. Due to the high rate constant of this reaction, detectable amounts of conversion can be achieved in time intervals in the sub-millisecond range. Chemical reaction data for the hydrolysis of phenyl chloroacetate showed a mixing dead time of 100 µs. Thus, the feasibility of initiating and quenching reactions with time intervals down to 110 µs in the micromixer could be proven.

3.5 Collision of High-Energy Substreams for Spraying/Atomizing

3.5.1 Collision of Three Substreams in a Microjet Reactor

One example of this flow configuration refers to a mixing device, termed microjet reactor, consisting of a spherical mixing chamber the center of which forms the focal point of the directions of three triangularly arranged small openings, manufactured by means of ultraprecision machining [13]. The operational principle of this device is due to the combined action of hydrodynamic cavitation and water jet cutting technique. The mixing concept is based on early descriptions of the use of cavitation, induced by ultrasonic waves, for nano-scale particle generation of uniform material properties [14, 24]. Thereby, 3 nm small iron particles were generated on a silica carrier. On the basis of these results, an apparatus was developed, termed a Microfluidizer [25], patented, and is now commercially available from the Freudenberg company in Germany. Main fields of applications refer to the ho-

Fig. 3-7. Microjet reactor using three filigree nozzles for high-energy collision of fluid streams [13].

mogenization of emulsions as well as the generation of piezoelectric and supraconducting materials.

Through the openings of the microjet reactor fabricated by Synthesechemie (Germany), three liquid streams enter at high velocity and jointly collide in the center (see Figure 3-7). One field of application refers to the crystallization of small particles using the extremely high energies released. These energies are so high that any contact of a stream with a wall of the mixing chamber, e.g. as a consequence of dejustification, would remove construction material. In order to avoid such damage, large ceramic spheres are inserted in the mixing chamber.

3.6 Injection of Many Small Substreams of One Component into a Main Stream of Another Component

3.6.1 Injection of Multiple Microjets

Mixing Device with Sieve-Like Bottom Plate
Researchers at the Mechanical Engineering Research Laboratory in Tsuchuira (Japan) and of the MESA Research Institute in Twente (Netherlands) developed a micromixer with a rectangular mixing chamber which contains a sieve-like bottom plate [15]. This plate consists of a large number of regular, square-like microholes which are arranged in several parallel rows (see Figure 3-8).

A liquid stream is guided through the mixing chamber, while another liquid is injected into this chamber through the microholes. Thereby, the second liquid stream is split into

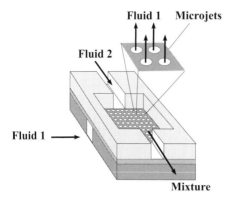

Fig. 3-8. Schematic representation of a micromixer based on the injection of multiple microjets into a mixing chamber via a sieve-like structure with a large number of regular holes [15].

many small jets, called "micro-plumes". By these means the contact surface between the two liquids is significantly increased.

The distance between the holes was fixed at 100 μm based on fabrication issues, while the hole length and width were varied from 10 to 30 μm. These dimensions determine the correspondingly short diffusion paths between the jets and the surrounding homogeneous liquid.

Calculations of Mixing Efficiency
Numerical analysis of the diffusional flux was used to demonstrate theoretically the feasibility of the mixing approach. It was shown that, as to be expected, mixing using the injection of a multitude of micro-jets into a fluid stream is much faster than mixing by one large single point injection, i.e. mixing in a configuration corresponding to a mixing tee. By means of multiple microjets, mixing is improved for all flow rates, with higher flow rates usually accelerating the mixing process.

The results of this analysis were taken into account in selecting a favorable design of the mixing device. A mixing chamber with dimensions of 2.2 mm x 2 mm x 0.33 mm supported by a sieve-like plate with 400 microholes of 15 μm x 15 μm size was chosen. Typical flow rates of this device were in the range 0.5 to 8 μl/s.

Microfabrication
Fabrication issues were not only addressed to the realization of the mixing component, but moreover also had to cope with the integration of the micromixer in a micro liquid handling system, containing dosing systems and detectors as well. In this context, bulk micro machining of (100) silicon wafers was chosen as a simple fabrication technology suitable for integration of several components on one chip.

The sieve structure and mixing chamber were etched from both sides in a silicon wafer. This wafer was covered on both sides with Pyrex glass wafers by anodic bonding. One glass wafer contained at the outside a light reflecting silicon layer of the same size as the mixing chamber. This few microns thin layer was deposited by thin film technology and structured by reactive ion etching. The liquid samples were introduced via in- and outlet ports manufactured by ultrasonic drilling in the top glass wafer.

Determination of Mixing Efficiency by Fluorescence and VIS Absorption
The experimental work included flow visualization for a qualitative analysis of the mixing process. A fluorescent dye was injected into the stream flowing to the sieve-like structure. The changes in fluorescence intensity induced after jet formation were monitored by imaging the diffusional mixing in the mixing chamber using a fluorescence microscope. These measurements proved the formation of a homogeneous mixture within a few seconds.

In addition, VIS absorption measurements using an optical fiber setup were performed directly in the mixing chamber. The optical path, and hence the signal-to-noise ratio, could be strongly increased by passing the light through the full length of the flat chamber. Since this could not be achieved by a simple transmission measurement, the authors used multiple reflections of the light beam between the bottom plate of the mixing chamber and the reflecting silicon layer. The light beam enters and leaves the mixing chamber by reflection at a 90° angle on adjacent chamber walls. This configuration allowed a pair of optical fibers, connected to the laser and the photo detector, to be placed on top of the mixer.

The VIS absorption measurements proved complete mixing at a flow rate of about 1 μl/s within a mixing time of about 1 s, which is in accordance with the microscopic analysis.

3.7 Manifold Splitting and Recombination of a Stream Consisting of Two Fluid Lamellae of Both Components

3.7.1 Multiple Flow Splitting and Recombination Combined with Channel Reshaping

While the two concepts presented in Sections 3.4 – 3.6 achieved mixing in one step, a number of approaches were based on multiple repetition of the mixing process in many mixing elements, literally forming a mixer cascade similar to macroscopic static mixers. In all these cascades a bilayered fluid system is formed at the beginning, which is split in each following step into more and more thin layers.

Mixer Cascade Utilizing Splitting–Reshaping–Recombination
IMM was among the first who introduced concepts for mixer cascades performing multiple division of fluid layers in 1994 and 1995 [7, 8, 26]. In these cases, this was achieved by a repeated sequence of splitting–reshaping–recombination processes. Thereby, the layer thickness of the two fluids to be mixed was decreased, enhancing the mixing process. The mixer was constructed of two interconnected plates which both contained microstructures. Each microstructure was an array of two parallel-operated cascades.

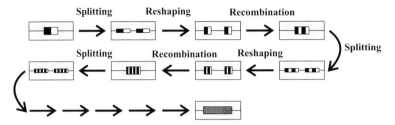

Fig. 3-9. Mixing mechanism based on a splitting–reshaping–recombination mechanism.

For reasons of simplicity, the flow in only one of these cascades is described in the following (see Figure 3-9). Two fluid layers are brought into contact in a channel in a tee-like configuration at the start of the cascade. Then, the layers are split perpendicular to their interface (1) by separating the channel in one upper and lower channel. At the beginning of the latter channels the height is reduced to half compared with the original geometry, while the channel width is kept equal. Along the flow axis, the channel width and height are simultaneously varied, keeping the cross-section of the channel constant (2). Reshaped channels are brought back to the same height level (3) and are recombined (4), re-establishing the original channel geometry. Within one splitting–reshaping–recombination cycle, the number of fluid layers is doubled, while the lamellae are bisected, resulting in a faster mixing process. By repetition of this cycle (5, 6, 7, 8) in the mixing cascade, the same improvements can be achieved again.

Microfabrication
The central element of the micromixer is the channel structure between positions 2 and 3 (see Figure 3-9). Channel depth and width are continuously and simultaneously varied between 70 and 140 µm. Therefore, a fabrication technique capable of 3D structuring has to be chosen. Since this excludes the use of standard lithographic processes, a variant of the LIGA process, the so-called Laser-LIGA, using an excimer laser was applied. Thereby, the channel depth was changed by varying the number of photon pulses.

The Laser-LIGA process basically consists of three steps, namely lithography, electroforming and molding. Details of microstructures, corresponding to this fabrication course, are illustrated in Figure 3-10 [27]. First, a cast polymer film is structured by excimer laser ablation, and subsequently, this master structure is filled with metal by electroforming. A sufficiently thick metal replica is obtained by several fabrication steps. After separation from the polymeric film, this part can be used as a mold insert for injection molding, yielding exact polymeric replicas (see Figures 3.10 and 3.11).

Figure 3-11 shows the overall microstructure of the final polymeric product, which is composed of an array of two parallel cascades each consisting of five mixing units. An image of a single mixing unit is given in Figure 3-10c. For interconnection, hot embossing and laser welding were applied. Figure 3-12 shows the molten and compressed interface

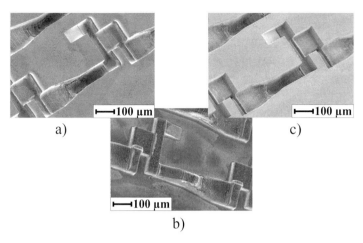

Fig. 3-10. Scanning electron micrographs of the polymeric master structure (PMMA) of the micromixer machined by excimer laser ablation (a), the nickel replica by electroforming (b) and an injection molded PMMA structure (c) [27].

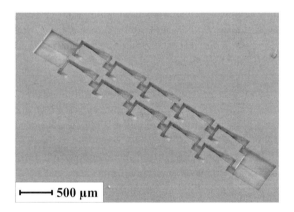

Fig. 3-11. Scanning electron micrograph of two parallel operating cascades of micromixing units [28].

achieved by hot embossing. Such joined polymeric layers demonstrated fluid tightness for liquid media.

Visual Observation of Mixing
The mixing process in this micromixer was analyzed by visual observation of the color changes after laminating two miscible liquids, one being intensively colored with a dye and the other consisting of pure water. A homogeneous color of the mixed solution was obtained after approximately 4 to 5 mixing elements corresponding to mixing times of the order of a few tens of milliseconds.

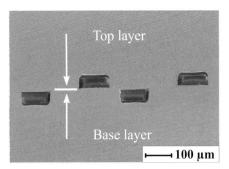

Fig. 3-12. Image of the interconnection zone of two polymeric layers containing microchannel structures of the micromixer [29].

3.7.2 Multiple Flow Splitting and Recombination Using Fork-Like Elements

Researchers at the TU Ilmenau (Germany) wanted to integrate a micromixer in a modular microfluidic system [30] which contained other microcomponents such as pumps, valves or sensors. The major issue in this development was the minimization of the system size by flexible integration of the components.

Mixer Cascade Utilizing Splitting–Rearrangement–Recombination
The micromixer design again was based on a concept for achieving multiple division of fluid layers in a mixer cascade. Here, a splitting–rearrangement–recombination sequence was chosen. In a series of fork-like elements the fluid was continuously laminated and separated (see Figure 3-13). This was achieved by combining the two outlet channels of each fork with the inlet channel of the next fork. Thereby, the number of fluid layers was doubled after passing one mixing element. However, different from most concepts, a triangular channel structure was chosen, so that no real lamellae are generated but fluid compartments with variable thickness and width.

Generally speaking, this concept is generically similar to bifurcation approaches. In a bifurcation unit, one channel is symmetrically split into two channels yielding an Y- or T-shaped system which resembles the fork unit. Whereas in the fork-like concept the two new generated channels are reunited, bifurcation continues to split each channel into two new ones. Thus, in the first case the channel number stays constant, whereas in the latter case the Y- or T-shaped structural units are continuously increasing. In order to achieve mixing, two bifurcation structures have to be combined. Thus, multiple splitting is performed prior to recombination. In the fork-like concept combined splitting and recombination processes are continuously repeated.

Microfabrication
Planar semiconductor techniques were employed for the microfabrication of the mixer. After deposition of a thermal oxide layer as etch stop and UV irradiation, a <100>-oriented

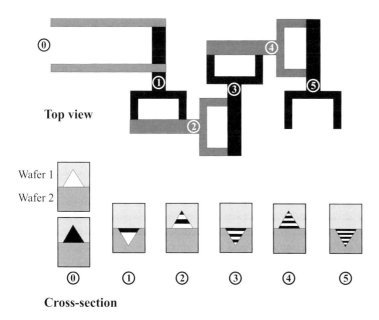

Top view

Cross-section

Fig. 3-13. Schematic representation of the fork-like micromixer containing five mixing elements performing multiple splitting–reshaping–recombination steps (top). For selected cross-sections idealized drawings of the ongoing fluid layer generation in the elements are given (bottom) [30].

p-doped silicon wafer was anisotropically etched using a concentrated KOH solution. Etching stopped on reaching the etch stop layer. After removal of the oxide surface layer, the wafer was modified in an oxygen plasma. The channels were covered with a top plate by low-temperature bonding at 400 °C and the microstructures were isolated by cutting. The fluid connection was achieved by insertion of stainless steel tubes connected to transparent polymeric tubes.

The overall size of the mixing forks was 500 µm x 500 µm x 720 µm. Each fork comprises triangular microchannels with widths of 150 to 200 µm at the top of the channel. A variation of the channels' cross-sections allowed the pressure drop to be adjusted in order to achieve a uniform flow distribution.

Visualization of Mixing Process/Dispersion Generation
The mixing process was characterized by means of the determination of a color change of an aqueous solution due to a pH shift. A homogeneously colored solution was observed after passage of about five mixing elements.

The mixer was used not only for mixing of miscible fluids, but also for multiphase contacting. A segmented flow of gas bubbles (air) with a diameter of 200–500 µm and liquid segments (water) was observed. An increase in the number of mixing elements from 5 to 10 resulted in a foam having even smaller bubble sizes below 100 µm. Similar findings were made by contacting a highly viscous oil with air. For these dispersions only a slow bubble growth could be detected showing long-term stability of up to three weeks.

In addition, contacting of non-miscible fluids, oil and water, in a mixer with 16–20 mixing elements resulted in emulsion formation. The non-transparent emulsion contained water droplets of a few 100 µm diameter in an oil phase which was stable for more than 70 days. Using mixers with smaller numbers of mixing elements led to alternating water and oil segments which, after leaving the outlet tube, collapsed to large water droplets of about 500 µm diameter.

3.7.3 Multiple Flow Splitting and Recombination Using a Separation Plate

A micromixer, designed by a research group at Danfoss and the Mikroelektronik Centret in Copenhagen (Denmark), was an integral part of a micromachined chemical analysis system including an industrial chemical sensor [9].

Mixer Cascade Utilizing Splitting–Recombination
The micromixer performs multiple splitting and recombination of thin liquid layers, thereby decreasing the layer thickness. As a further development of earlier splitting–recombination concepts, here the lamination is actively supported by the use of a separation plate (see Figure 3-14). Thus, the flows stack on each other without forming a fluid/fluid interface which may disturb lamination. After stacking, the contact between the two liquids is achieved through multiple slit-shaped openings in the separation plate. Splitting is achieved by sim-

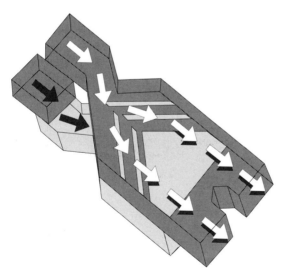

Fig. 3-14. One of several mixing units of a static micromixer with multiple slit-shaped injection openings. Such mixing units can be connected in series to improve mixing quality by multiple flow splitting and recombining [9].

ply dividing the upper channel into two channels of half width, i.e. resembling an inverse Y-tube.

Microfabrication
The micromixer is assembled from a silicon and a glass wafer connected by anodic bonding. One channel structure is etched into glass and the other into silicon. In the region where the channels overlap, they are separated by the separation plate defined by an etchstop layer. The channel covered by this structured plate was generated by underetching in the <100> direction through slits in the plate. The maximum width of the channels is 300 μm and the depth is 30 μm. The width of the channels for the inlet branches of the micromixer units is about 150 μm. The thickness of the structured plate for separating the channels in the glass and the silicon wafer is 5 μm and the slit width is 15 μm.

Theoretical Analysis of Flow Multilamination
The mixer design was based on a detailed theoretical analysis of flow and diffusion phenomena in the laminar regime for idealized channel geometry. The importance of the incorporation of theoretical calculations in the design and testing phase exemplarily was shown by comparing two generic multilayer architectures. The time evolution of so-called asymmetric and symmetric concentration profiles was first compared for stationary conditions, i.e. without any flow.

In the symmetric case, three inner layers of equal thickness, termed B−A−B, are surrounded by two layers A of half thickness, yielding an overall layer sequence A−B−A−B−A. In the asymmetric case all layers A and B have equal thickness, but the outer layers belong to two different fluids. Thus, the sequence refers to A−B−A−B. The time required to achieve a homogeneous concentration profile, i.e. the mixing time, is much longer for the asymmetric lamination.

In addition, dynamic studies were performed including moving fluids. Generally, similar results were obtained. Different from the stationary case, the width of the inner lamellae is decreased relative to the outer layers due to the parabolic flow profile.

Fluid dynamic simulations revealed the influence of cross-flow through the slits in the separation plate which were introduced only for reasons of microfabrication, i.e. underetching the plate to connect in- and outlet microchannels. It could be shown that the present design suffered from the problem that about 30 % of the total flow was fed through the slits, leading to uneven lamination. Further analysis demonstrated that this effect is reduced either using thinner slits, i.e. increasing the pressure loss, or combining several mixing elements in one cascade.

Flow Visualization of Mixing Process
The mixing process was observed by flow visualization using light microscopy. A pH reaction between phenol red and an acid yielded a color change. Thereby, it was possible to resolve zones with incomplete mixing in the neighborhood of the contacting zone, in particular those referring to the uneven lamination caused by the slits. A homogeneous mixture was formed in the outlet channel. Complete mixing in the whole cross-section of the

outlet channel was obtained in 100–300 ms. The flow range applied was 0.01–0.1 µl/s, which corresponds to a highly viscous flow with a Reynolds number less than 1. The mixing times measured were not dependent on the flow rates.

3.7.4 Multiple Flow Splitting and Recombination Using a Ramp-Like Channel Architecture

High-Throughput Mixer
High throughput was a key issue in a micromixer development of the Institut für Mikrotechnik Mainz (Germany) [12, 31, 32]. Furthermore, the structuring process had to be designed in order to permit small-series fabrication.

The device, termed a caterpillar mixer, consists of one main channel formed by assembling two microstructured plates. The channel geometry is characterized by a complex ramp-like architecture (see Figure 3-15). Thereby, the fluid is constantly moved up and down and, simultaneously, split and recombined.

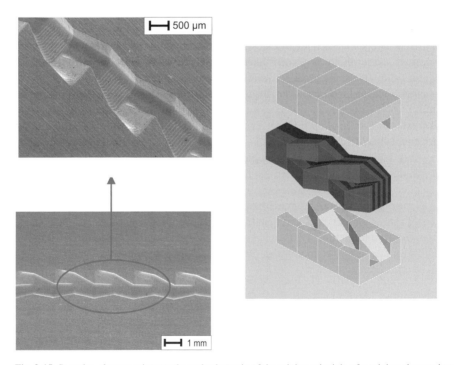

Fig. 3-15. Scanning electron micrographs and schematic of the mixing principle of a mixing element based on ramp-like architecture. The idealized flow leading to a multilayered system is also indicated [12, 31, 32].

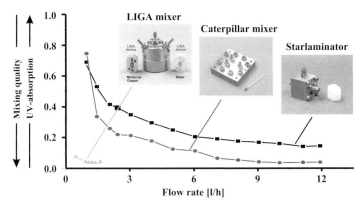

Fig. 3-16. Schematic of the dependence of the mixing quality on the volume flow for a ramp-like micromixer. For comparison, the corresponding curves of the interdigital micromixer and star laminator are also drawn [32].

Different from many other variants, such as the concepts discussed in Sections 3.6.1 to 3.6.3, no additional fluid connecting channels are needed between a splitting and recombining step. In addition, separation of flows can be achieved without the need to divide them by walls, in other words, necessitating a multitude of single channels. Therefore, large throughputs at a given pressure loss are achievable and also maintenance, e.g. cleaning, is facilitated.

Microfabrication
The mixer plates were fabricated by micromilling. The channel width and depth are both 1200 µm. The length of one unit cell consisting of two ramps amounts to 2.4 mm. The mixer actually realized contained six unit cells, arranged in series, corresponding to a total length of 19.2 mm. The incoming fluids have a thickness of 600 µm, which is reduced to about 2 µm after the eight split–recombine cycles.

Analysis of Mixing Quality and Droplet Size Distribution
The mixing quality was characterized by the method of competing reactions (see Figure 3-16 and also Section 3.8.1). The quality achieved at high volume flows, ranging from 7 to 12 l/h, was close to that of the interdigital mixer (see Section 3.8.1) which can only be operated at much lower flow rates. Thus, the feasibility of a high-throughput concept with excellent mixing quality could be demonstrated.

Furthermore, experiments concerning multiphase contacting were performed. Emulsion formation using the model system silicone oil/water was possible for a large range of operating conditions. Droplet sizes of the order of 40 to 60 µm were found for the highest flow rate of 12 l/h achievable with standard piston pumps. Hence the droplets are about one to two magnitudes larger than emulsions generated by the interdigital mixer (see Section 3.8.1).

3.8   Injection of Many Substreams of Both Components

3.8.1   Multilamination of Fluid Layers in an Interdigital Channel Configuration

Versatile Micromixer Concept Based on Flow Multilamination

A micromixing device developed by the Institut für Mikrotechnik Mainz (Germany) uses flow multilamination of fluid lamellae with subsequent diffusional mixing [12, 16, 20, 22, 33]. Both fluids are split into many substreams which are dispersed, yielding a multilayered system with a typical layer thickness of a few tens micrometers.

The micromixer has been realized in a variety of materials and with different designs. A huge number of experiments concerning mixing of miscible and non-miscible fluids have been performed. This choice of materials and designs as well as the broad experimental know-how stimulated interest from industry and institutes so that the micromixer has become a commercially successful microreactor product available in a small-series production. In this context, a larger chapter is needed for the description of the device and the presentation of the results.

The design of the micromixer is shown in Figure 3-17. The fluids to be mixed are introduced into the mixing element as two counter-flows which stream into an interdigital channel configuration with corrugated walls. Typical values of the channel widths are e.g. 25 or 40 μm. By means of the slit-shaped interdigital channels a periodic configuration is generated consisting of the flow lamellae of the two fluids to be mixed. The laminated flow leaves the device perpendicular to the direction of the feed flows and, because of the small thickness of the lamellae, fast mixing takes place through diffusion (see Figure 3-18). The corrugated shape of the channel walls serves the purpose of increasing the contact surface of the laminated streams and improving the mechanical stability of the separating walls.

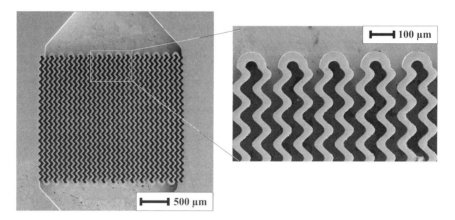

Fig. 3-17. Scanning electron micrographs of a mixing element based on multilamination of thin fluid layers. The device consists of 2 x 15 interdigitated microchannels with corrugated walls fabricated by means of LIGA technology [16].

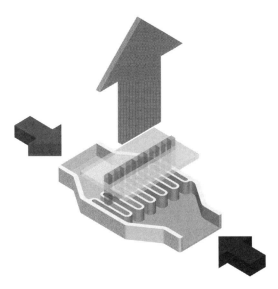

Fig. 3-18. Multilamination of streams in an interdigital channel configuration with corrugated walls leading to fast mixing through diffusion [16].

The optimum operating conditions of such a device depend on the height and the length of the corrugated interdigital channels as well as on the properties of the fluids to be mixed such as viscosity, flow rate, pressure, etc. A very simple adjustment of the flow and mixing conditions is obtainable when the fluid is withdrawn not over the total surface of the interdigital structure, but through a slit which forces interpenetration within a defined contact zone.

Microfabrication

Mixing elements made of various metals were realized in the frame of the LIGA process by deep X-ray lithography and electroforming. Nickel and silver elements have been favorably applied for commercial devices. In particular, silver proved to be resistant to a number of solvents, including acidic and basic media. Furthermore, the fabrication of polymeric elements by injection molding using LIGA-made mold inserts is under way [29]. For instance, PMMA mixers are suitable for handling of aqueous media. Mixers made of engineering plastics can even be operated with many organic solvents.

Furthermore, mixing elements made of materials with extremely high corrosion resistance are required. μ-EDM techniques were employed to fabricate titanium mixing elements using electrodes made by thin-wire erosion [12, 32]. Another approach utilizes the high chemical resistance of glass-like surfaces. A silicon oxide layer is generated on a silicon mixing element, realized by advanced silicon etching, by a thermal diffusion process.

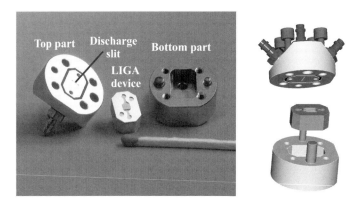

Fig. 3-19. Photographs of a micromixer consisting of a LIGA device with an interdigital channel structure and top and bottom parts [20].

Figure 3-19 shows the assembly of the mixing element into a housing. The element is inserted into a bottom part and covered with a top part which contains the two inlets of the feed streams and the outlet of the mixture. The discharge slit is also integrated in the cover part. Top and bottom parts are available in a wide variety of materials. To date, the standard housing material is a high-alloy stainless steel. Plastics and titanium have also been employed. In particular, systems composed of titanium housings with titanium or oxidized silicon mixing devices demonstrated a high corrosion stability, e.g. if highly concentrated halide containing aqueous solutions were applied.

Mixing of Miscible Liquids
Interdigital micromixing systems have been tested extensively with miscible fluids. A chemical method based on competing reactions was used for characterization of mixing efficiency. Two different feed fluids react after mixing via two different pathways, a fast one and an ultrafast one, resulting in different colors of the reaction product [34]. In case of the fast reaction, the reaction product is colored whereas for the ultrafast reaction, which dominates at ultrafast mixing, a colorless solution is obtained.

$$5\ I^{\ominus} + IO_3^{\ominus} + 6H^{\oplus} \longrightarrow 3\ I_2 + 3\ H_2O$$

The measurements showed that complete mixing is obtained a few millimeters downstream of the contacting zone of the laminated flows, which corresponds to a mixing time less than some tens of milliseconds. The flow rate applied in these experiments ranged from 10–1000 ml/h; the pressure losses were between 10 and 1000 mbar. For qualitative comparison, the feed fluids mentioned above have also been mixed using a mixing tee with corresponding sizes of feed and withdrawal channels as well as a heavily stirred mixing vessel. It turned out that such standard devices could not compete with the mixing speed of the LIGA micromixer (see Figure 3-20). In particular, it has to be emphasized that the

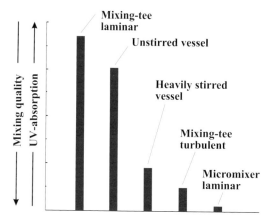

Fig. 3-20. Comparison of mixing quality of the interdigital micromixer and standard mixing devices, laboratory stirrers and mixing tees. Mixing tees and micromixer were operated at the same volume flow [16].

micromixer gave a better mixing performance than the mixing tee operating in the turbulent regime.

The test reaction utilizing the two parallel reactions was, in particular, useful in establishing a quantitative measure of the mixing performance of various micromixers. First experiments focused on the influence of the variation of structural details of the mixing element and the discharge slit on the mixing quality. It was demonstrated that, if the fluid is not withdrawn over the total area of the mixing element, i.e. restricting the slit width to a central zone, mixing performance is significantly improved. This is caused by a deeper penetration of fluid lamellae. A similar mixing improvement was achieved by using mixing elements with smaller microchannels, e.g. having a channel width of 25 μm instead of 40 μm, which results in a decrease in lamellae thickness.

In addition to these geometric influences, the impact of the operating conditions was investigated. Figure 3-21 shows that over a large range of volume flows a nearly constant mixing quality is achieved, as to be expected for a micromixer operating in the laminar regime. At volume flows smaller than 150 ml/h a large increase of absorption is observed. Limitations in the applicability of the test reaction demand a careful interpretation of the experimental results.

Emulsion Generation of Immiscible Liquids
Apart from homogeneous mixing, the utilization of interdigital micromixing systems is particularly interesting for generating emulsions [22]. A model system, a low-viscous silicone oil and water, was chosen in order to characterize the quality of the emulsions generated. Suitable parameters for comparison of different operating regimes and types of micromixers are the average droplet size and the mean deviation from this average.

These parameters were related to structural details of the mixing device and the operating conditions. It was evident that monomodal distributions of droplets of one liquid in

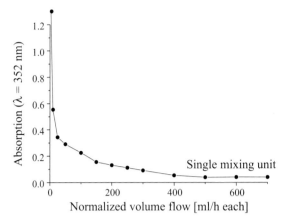

Fig. 3-21. UV absorption of the single mixing unit versus volume flow [16]. The UV absorption is inversely related to the mixing performance.

another can be obtained over a broad range of operating conditions (see Figures 3.22 and 3.23). The emulsification quality was strongly improved at higher volume flows, in contrast to the findings with homogeneous mixing. Smaller droplets were found and the standard deviation of the droplet size became narrower.

Thus, the energy introduced in this process regime has a huge impact on the dispersion process. A tentative explanation, based on the generation of periodic velocity gradients, for this has been given [22]. Assuming this theory, dispersion should be improved also by operating the two liquids at different flow rates. Such an effect could be proven because

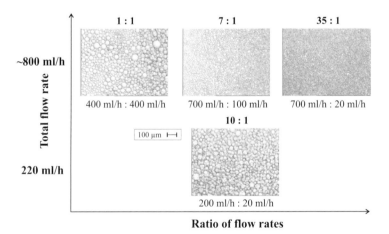

Fig. 3-22. Phenomenological analysis of the dependence of the average droplet size, generated in the interdigital micromixer, on the total flow and the ratio of the flow rates of silicone oil and water [22].

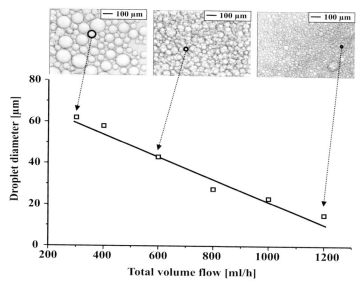

Fig. 3-23. Droplet diameter of silicone oil–water emulsions, generated in the interdigital micromixer, as a function of the total flow for an 1:1 ratio of the flow rates of the two liquids [22].

very small droplets down to 4 µm resulted if flow rates as high as 700 ml/h:20 ml/h, i.e. a ratio of 35:1, were applied (see Figure 3-22).

In addition to the influence of the operating conditions, structural details such as the channel width were varied. The use of smaller channels led to smaller average droplet sizes, similar to the results concerning miscible liquids (see Figure 3-24).

Although within the investigations no detailed analysis of the stability of the emulsions was performed, it was noticed that some emulsions turned out to be stable over a period of at least some months, even without the use of any standard emulsifier.

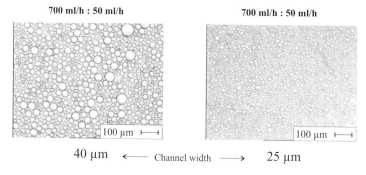

Fig. 3-24. Phenomenological analysis of the dependence of the average droplet size of silicone oil–water emulsions, generated in the interdigital micromixer, on the microchannel width [22].

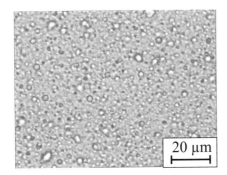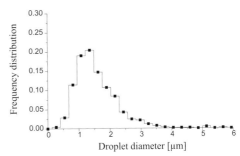

Fig. 3-25. Photographic image of a cream, generated in the interdigital micromixer, revealing droplets smaller than 1 μm (left). Frequency distribution of the corresponding droplet diameter (right) [18].

Industrial Applications of Emulsions

First industrial applications of the use of micromixers for the generation of emulsions and semi-solid creams have been reported [18]. Using a commercial mixture of waxes, water, drugs and detergents, a cream can be generated which is semi-solid at room temperature, but liquefies at about 60 °C. The aim of the studies was to control the viscosity behavior of the cream, by setting the droplet size, to allow a controlled drug delivery when applied on human skin. It was found that using the micromixer, creams with an average droplet size of 1 μm could be generated (see Figure 3-25). The smallest mean droplet sizes observed were about 800 nm and the largest did not exceed 4 μm, i.e. an extremely narrow distribution could be achieved.

Gas/Liquid Contacting

Gas/liquid dispersions were also investigated using the micromixer [23]. By simply contacting water and argon, a bubble flow of rather undefined bubble sizes resulted. Adjusting viscosity and surface tension by addition of glycerol and a detergent led to a uniform distribution of bubbles flowing in an ordered packing similar to the structure of a cubic-oriented crystal (see Figure 3-26). Bubble sizes down to 100 μm have been found. Details of the flow pattern and its dependence on the operating conditions have been described [23]. Improved integrated systems based on the LIGA micromixing elements have meanwhile been realized and are currently being tested. A detailed description of the devices and first results are summarized in Section 8.2.5.

Industrial Applications of Gas/Liquid Dispersions

Based on these results, the performance of homogeneously catalyzed multiphase reactions including two liquid and one gas phases was investigated [12, 32]. Liquid/liquid and gas/liquid contacting was performed by a serial connection of two micromixers. The aim of the study was the optimization of an existing process by operating in the explosive regime. Over a period of several months a safe handling at both high pressures and temperatures

Fig. 3-26. Flowing dispersion of argon bubbles in an aqueous solution generated by means of the interdigital micromixer [23].

was achieved. Although the conversions measured were only similar to the established process, benefits in terms of the cost function of the equipment could be clearly shown. Safety instrumentation of the large-scale reactor causes significant additional costs, whereas such instrumentation is not needed when using the micromixers.

Parallelization Concepts: Numbering-up for Increase in Throughput
Several attempts have been made in order to improve the throughput of the mixer by numbering-up of several mixing elements into so-called mixer arrays.

In an early approach, ten mixing elements were assembled in parallel on a circle (see Figure 3-27). Feeding was performed from an outer circular channel which had branches to the mixing elements. The other fluid enters the mixer from the center of the device from which it is distributed to the mixing elements. Mixing is performed within a dodecagon-shaped channel of variable width (several hundred μm) in the top plate in analogy with the discharge slit of the single mixer device.

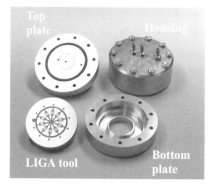

Fig. 3-27. Photograph of the mixer array consisting of top and bottom housing plate and LIGA device [16].

71

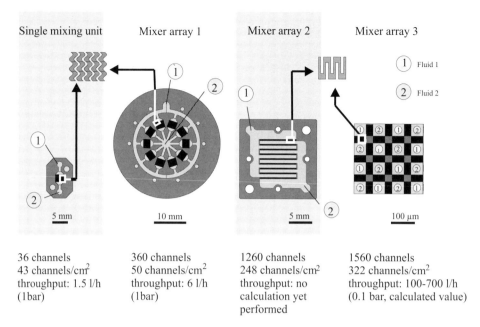

Single mixing unit	Mixer array 1	Mixer array 2	Mixer array 3

36 channels
43 channels/cm^2
throughput: 1.5 l/h
(1 bar)

360 channels
50 channels/cm^2
throughput: 6 l/h
(1 bar)

1260 channels
248 channels/cm^2
throughput: no
calculation yet
performed

1560 channels
322 channels/cm^2
throughput: 100-700 l/h
(0.1 bar, calculated value)

Fig. 3-28. Comparison of number density of microchannels per device and unit volume for different concepts of parallelization of microchannels.

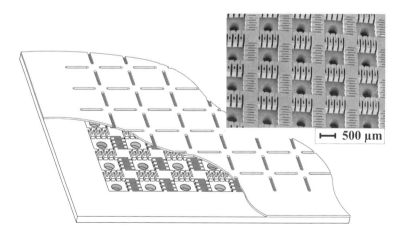

Fig. 3-29. Schematic of a mixer array comprising honeycomb-like arranged feed regions separated by rows of interdigitated microchannels (left). Scanning electron micrograph of details of the mixer array, revealing several interdigital zones and feed regions (right) [32].

By this type of parallelization, the throughput could be enhanced by about a factor of six compared with the single device. The ideal factor of ten was not achieved due to additional pressure losses induced by the feed lines needed for flow distribution.

Concerning compactness issues, the circular arrangement still is not the optimal solution. IMM proposed a parallel assembly of long rows of interdigitated microchannels (see Figure 3-28). Compared with the single device and the mixer array having 32 and 320 microchannels, respectively, the newly developed device contains 1260 channels. Elongated discharge slits serve as mixing chambers in this concept.

Finally, the highest degree of integration can be achieved by a honeycomb-like arrangement of square-shaped feed regions separated by rows of interdigitated microchannels (see Figure 3-29) [32]. The feed regions contain boreholes which are connected to underlying feeding reservoirs. A honeycomb-like slit system is positioned above the rows. Currently, one silicon wafer with 1536 feed regions is fabricated by means of Advanced Silicon Etching (ASE). A flow of up to 1000 l/h at reasonable pressure drops below 1 bar was estimated from simulations.

3.8.2 Vertical Multilamination of Fluid Layers Using a V-type Nozzle Array

Researchers at the Mikroelektronik Centret in Copenhagen (Denmark) designed a micromixer for the performance of chemical and biochemical analysis including the mixing of suspensions [2].

Flow Multilamination in a Nozzle-Type Mixer
The micromixer comprises a minichannel with a larger chamber in its middle section where two rows consisting of U-shaped microstructures are embedded (see Figure 3-30).

The openings of these microstructures are arranged towards the outlet direction of the channel. The rows touch each other at one end and are tilted relative to the middle axis of the minichannel, i.e. the direction of the incoming flow. As a result, a V-shaped array is formed similar to a filter structure with regular pores (see Figure 3-31). In this context, the authors name the interstices and openings of the microstructures as "nozzle".

One fluid enters this V-type nozzle array from the rear and, by passing the interstices, is split into many substreams. The other fluid is fed through holes within the U-shaped microstructures from a reservoir located below the base plate of the chamber. After leaving the interstices and microstructures, the fluids come into contact as thin parallel oriented lamellae. Since microstructures of relatively high aspect ratio can be generated by the structuring technique chosen (see below) the authors use the term vertical lamination for the resulting flow arrangement. This term refers to a special generic layer configuration which has already been introduced and defined in an earlier publication [9].

An analysis of the influence of the mixer geometry on the multilayer formation as well as the reliability and applicability of the system is given. Calculations show that the length of the microstructure interstices as well as the openings are sufficient to allow for a fully developed flow profile of the incoming fluid stream before lamination is achieved. There-

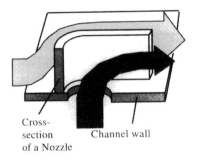

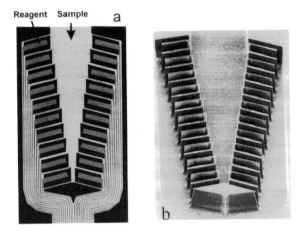

Fig. 3-30. Schematic of the mixing principle: lamination of two streams using nozzle structures (top). Top view of mixing element, displaying schematic and real images (bottom) [2].

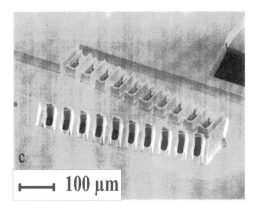

Fig. 3-31. Scanning electron micrograph of the mixing element with nozzle structures [2].

fore, only large reservoirs and openings, besides a small section with tiny microchannels, have to be passed. This feature is, in particular, important for applications in enzymology and biochemical analysis involving suspensions of larger objects such as cells in water. The clogging phenomena, usually associated when using suspensions, can be minimized with the design chosen.

Microfabrication

The mixer structure was fabricated by anisotropic reactive ion etching (RIE) of a silicon wafer in a plasma composed of SF_6 and O_2. The U-shaped microstructures and holes were etched from the front side with a double masking layer of oxide and resist. A structural depth of the channel system of 50 μm was realized. The interior of the microstructures was etched 25 μm in addition. In combination with backside etching using KOH, this resulted in a breakthrough, i.e. the formation of a hole. A sealing plate made of glass was bonded to the silicon wafer.

The fabrication route chosen allows a fast and inexpensive variation of structural details of one common design in order to flexibly adjust to the needs of different operating conditions. This flexibility is a common feature of all devices fabricated by means of parallel processing, wafer-based microtechnologies. For instance, an increase of sample throughput may be achieved by simply changing the number of U-shaped microstructures. Moreover, a process optimization with respect to lamination and minimization of dead volume can be performed by variation of the dimensions and position of these microstructures as well as the geometry of the channel.

Visualization of Mixing Process

The system performance was characterized by observation of the lamination process through the transparent top plate using a video camera. For flow visualization, the color change after mixing of a solution of hydrochloric acid and phenol red, a pH indicator, was monitored. The laminated layers of both solutions were clearly visible as thin dark and uncolored stripes. In the central part of the V-shaped nozzle array perfect lamination was achieved directly behind the microstructures. In contrast to that, lamination was incomplete at the outermost microstructures, obviously due to insufficient flow homogenization.

3.8.3 Multilamination Using a Stack of Platelets with Microchannels

Stacked Platelet Devices with Straight and Curved Channels

Researchers at the Forschungszentrum Karlsruhe (Germany) developed mixing devices with a stack-like architecture of microstructured platelets which comprise hundreds of microchannels in parallel [17, 35–37]. Since the platelets are alternately connected to two fluid reservoirs, a multilaminated fluid layer system is obtained after contacting of the numerous streams in a large mixing chamber.

Two concepts for connecting the reservoirs to the mixing chamber have been investigated and patented (see Figure 3-32). One design utilizes straight channels. Since the chan-

P-Mixer: curved channels   **V-Mixer: straight channels**

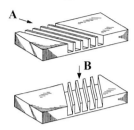

Fig. 3-32. Two platelet designs based on straight or curved channels. The platelets are arranged to a stack yielding a multilayered fluid system when fed with two fluids [38].

nel axis is tilted relative to the axis of the platelet and, consequently, of the mixing chamber, two separate inlet ports result which are connected to external feed reservoirs. The buildup of a pressure barrier at the inlet ports and the equal length of all channels on one platelet guarantee flow equipartition.

In addition, a second design was proposed using curved channels, termed P-mixer. The curvature of the channels was adjusted in such a way that the entrance streams of both fluids are directed parallel to the axis of the mixing chamber. A disadvantage of using curved channel systems is that they generally differ in length. In order to compensate this, the researchers in Karlsruhe used only a small curvature and, furthermore, adjusted the channels' length by suitable cutting of the inlet port region. Hereby, channels of nearly equal length could be achieved.

Microfabrication
The platelets of the Karlsruhe mixer were fabricated by means of mechanical surface shaping of metal tapes [39]. According to this fabrication process, typical channel depths from about 80 up to several hundred micrometers can be generated. Since the width of the remaining platelet base material is nearly negligible compared with this, the width of the corresponding lamellae should be only slightly higher. The platelets were made of various materials, e.g. stainless steel, hastelloy, and palladium.

Simulations of Mixing Efficiency
Simulations of the mixing behavior of the first concept based on straight channels predict a complete mixing of methane and oxygen within 30 µs at a gas velocity of 15 m/s. In addition, first industrial tests were conducted in order to evaluate the performance of the mixers for various reactions. Details of these tests have not been reported yet.

Experimental Determination of Mixing Efficiency
Hönicke and Zech of the TU Chemnitz (Germany) characterized the mixing efficiency of the micromixer with straight channels (V-mixer), manufactured at the Forschungszentrum

Karlsruhe (Germany) [38]. Channel height and width were set to 100 μm, respectively, channel length being 20 mm. 12 foils with 40 channels each were utilized.

The mixing efficiency was characterized in the mixing chamber by means of a 30 μm thin capillary, positioned by two motors in X-Y direction, which was connected to a quadrupole mass spectrometer. Thereby, a high spatial resolution was achieved. As test systems, argon/nitrogen and helium/nitrogen were applied at volume flows in the range of 75 to 300 l/h per passage, corresponding to gas velocities of about 4.3 to 17.4 m/s.

In first scouting experiments, the feasibility of the analytical method chosen was impressively demonstrated. For the argon/nitrogen system, measurements of the argon concentration close to the microchannel outlet (50 μm distance), i.e. at the start of the mixing

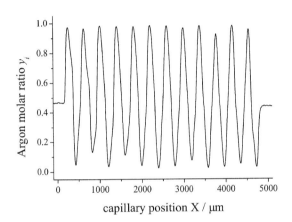

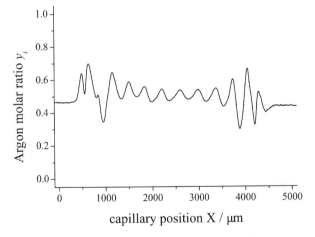

Fig. 3-33. Measurements of argon concentration for a multilaminated argon/nitrogen gas mixture at two locations in the mixing chamber of the stacked-platelet micromixer: 50 μm distance from injection (top), 550 μm distance from injection (bottom) [38].

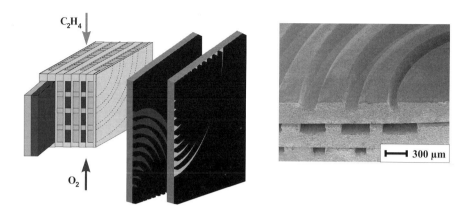

Fig. 3-34. Platelet design based on 90° curved channels. Channel width and length are adjusted so that an equal pressure loss results for all channels [40].

process, were carried out. A periodical concentration profile, changing from nearly pure to vanishing argon contents, was found, thereby indicating a very uniform multilayer formation which is not modified by interdiffusion of gas molecules. On the contrary, measurements at a more far position (550 μm distance from microchannel outlets) showed a much more averaged concentration profile, still being modulated by small periodical changes. Thus, gas composition clearly changed along the flow axis due to mixing by interdiffusion of the gases.

Determination of mixing efficiencies, based on mean deviations of concentrations from their final value, as a function of the mixing length evidenced that completed mixing of the argon/nitrogen system occured after only about 800 μm flow passage. This mixing length is in good accordance to the theoretical values derived by simulations, mentioned above. At higher gas velocities a longer mixing length was needed to achieve the same mixing efficiency as at small velocities.

A comparison of the results of the argon/nitrogen to the helium/nitrogen system reveals the impact of their diffusion coefficients on the respective mixing efficiencies. Even fine details such as the locally enhanced mixing efficiency at the crossing point of fluid substreams could be accurately resolved by the detection technique applied.

Stacked Platelet Devices with Curved Channel Geometry
Researchers at IMM in Mainz developed another concept based on a much higher curvature, i.e. a 90° tilted array of channels (see Figure 3-34) [40]. The length and width of the microchannels are adjusted so that an equal pressure drop results, as proven by fluid dynamic simulation and experimental results.

3.8.4 Multilamination Using a Stack of Platelets with Star-Shaped Openings

Combination of High-Throughput and Mass Fabrication Capability
The Institut für Mikrotechnik Mainz (Germany) developed a micromixer for high-through-put applications which had to be accessible by mass fabrication techniques [31].

The design was based on a stack of platelets with star-shaped openings (see Figures 3.35 and 3.36). Two types of star-shaped platelets were used to allow a separate feeding of alternating assembled platelets. The first type is fed via platelet openings connected to a fluid surrounding the stack. The second fluid is introduced via boreholes in the outer parts of the stars in the other type of platelet. The central openings of both types of platelets form a large mixing channel when alternately arranged as a stack. The stack is embedded in a housing and tightened by pressure forces applied on the stack by the top part of the housing.

Microfabrication
The platelets can be fabricated by either thin-wire erosion or punching. Since these cutting processes can be performed simultaneously using a stack of many platelets, a large number of pieces can be produced inexpensively in a short time. This is, in particular, valid when

Fig. 3-35. Photograph of a disassembled micromixer with star-shaped openings [29].

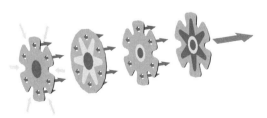

Fig. 3-36. Schematic of the sequence of platelets which reveals the flow of the fluids, finally leading to multilamination.

79

choosing the latter method. The housing was made by conventional precision engineering. Platelets of 20 and 50 µm thickness were generated. The complete stack was composed of 400 platelets.

Determination of Mixing Quality and Droplet Size Distribution
First measurements of the mixing quality (see Figure 3-37) based on the method of competing reactions revealed a performance inferior to that of the interdigital mixer (see Section 3.8.1) and the caterpillar mixer (see Section 3.7.4), but still significantly better compared with simple mixing tees and batch-type mixing using laboratory stirrers (see Section 3.8.1). Since for all three micromixers lamellae thickness of the order of a few tens of micrometers is expected, this result is surprising and needs further technical investigations. Therefore, work is currently aimed at improving the surface quality of the platelets in order to decrease leakage flows between the compressed platelets.

Emulsion generation based on the model system silicone oil/water is feasible. The dispersions generated are tentatively larger in droplet size compared with results with the interdigital mixer (see Section 3.7.1).

3.9 Decrease of Diffusion Path Perpendicular to the Flow Direction by Increase of Flow Velocity

3.9.1 Decrease of Layer Thickness by Hydrodynamic Focusing

The group of Austin at Princeton University in New Jersey (U.S.A.) developed a micromixer capable of diffusive mixing in extremely short times, e.g. down to 10 µs [19]. Information on this time scale has to be gained, e.g. when studying reaction kinetics of fast reactions.

Ultrafast Mixing via Hydrodynamic Focusing
Ultrafast diffusive mixing can only be achieved by an extreme reduction of the fluid layer thickness being not more than several tens of nanometers. Microfabricated devices alone are not adequate to generate such thin lamellae. Therefore, a further compression of the micrometer thin lamellae, generated in a microstructure, has to be performed by hydrodynamic means. This is achievable by contacting two fluids of extremely different volume flow. Due to volume conservation, the fluid of higher flow tends to form a thick lamella, while an ultrathin lamella of the other fluid results. Since the final thin lamella thickness is achieved after only a short entrance flow region, the whole process is termed "hydrodynamic focusing".

In order to analyze mixing phenomena occurring very close to the first fluid contact, i.e. directly behind the mixing element, a planar design completely covered by a transparent plate was chosen. This permits characterization without any dead times, hence permitting observation of the entire mixing process.

The mixing element contains a cross-like structure with four rectangular channels, each 10 µm deep and wide (see Figure 3-37). One fluid, the so-called inlet flow, is introduced

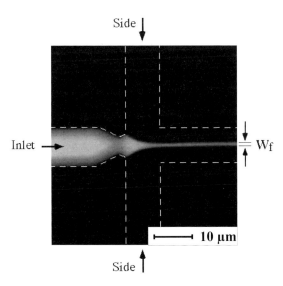

Fig. 3-37. Mixing element with cross-like structure [19].

through a channel which narrows to 2 μm at the intersection. The other fluid is fed through two channels both perpendicular to the first channel and facing each other at the intersection. This so-called side flow squeezes the other fluid into a thin stream that exits the intersection through the fourth channel of the cross.

Microfabrication
The mixers were realized by means of silicon micromachining using standard photolithographic techniques and a chlorine reactive ion etching process. The structured silicon wafers were sealed by a thin coating of cured silicone rubber. This silicone layer could be withdrawn after use allowing cleaning of the mixer. After repetition of coating, experiments could be continued using the cleaned structure. Four holes, for fluid connection of inlet and side flows as well as of the mixture to external sample reservoirs, were drilled into the backside of the wafer. Sample dosing was achieved without using any pump. Instead, the liquids were fed by applying pressure, controlled by regulating the incoming flow of nitrogen gas, on the head of each reservoir. The finite precision of pressure control directly influenced the minimum thickness of lamellae achievable.

Variation of Mixing Time by Adjustment of Feed Pressures
The focusing width was controlled by varying the volume flows of the side and inlet flows. Widths as small as 50 nm were measured, yielding a nearly instantaneous interdiffusion of the side flow into the inlet flow. However, it should be mentioned that this is not true for the dispersion of the inlet flow within the overall mixing volume, i.e. the fluid of the side flow has to serve as an excess reagent.

After mixing, the time evolution of any subsequent reaction can be spatially separated, and the resolution is only determined by the flow velocity. A resolution better than a microsecond per micrometer was achieved since laminar flow was even maintained at high velocities, due to the small size of the channels. A further consequence of the small mixer dimensions is low sample consumption. The volume flows of the focused reactant stream are typically on the scale of nanoliters per second, which is more than three orders of magnitude lower than to typical rates needed for turbulent mixers. This is, in particular, important for applications in biochemistry, usually demanding the consumption of expensive samples.

The authors found that hydrodynamic focusing was achievable in the mixer over a wide range of volume flow ratios, expressed as the ratio α of the corresponding pressures, defined as follows, where P_s is the pressure of the side flow channel, P_i refers to that of the inlet flow channel:

$$\alpha = P_s / P_i \qquad (3.2)$$

The minimum and maximum values of α, limiting the range of focussing, were determined to be $\alpha_{min} = 0.48$ and $\alpha_{max} = 1.28$. Increasing α results in a narrowing of the focused stream. Below the lower limit, the inlet flow enters the side channels, instead being directed to the outlet channel. Above the upper limit, the same phenomena occur for the side flow, hence streaming into the inlet flow channel. Both flow reversals were independent of the overall pressures P_s and P_i.

The experimental data for α were compared with theoretical values calculated by means of analogy considerations to electric flow (Ohm's law). Simple circuit models based on a network of resistors were applied to simulate the cross-type configuration chosen. It could be shown that the calculated values for α_{min} and α_{max} were in excellent agreement with the experimental data.

Visualization of Mixing Process by Fluorescence
The mixing process was visualized by epifluorescence and confocal microscopy images (see Figure 3-38). A bright inlet flow, labeled by a fluorescent dye, was mixed with nonfluorescent buffer side flows. The light emission of excited molecules of a focused fluorescein solution was quenched by a solution containing iodide ions. Thus, the time evolution of iodide diffusion into the fluorescein solution was revealed as a change in the fluorescence intensity.

Lamellae of the fluorescein solution larger than 1 µm could be directly imaged. In addition, a nonimaging approach, based on a fluorescence technique, was applied to analyze even smaller lamellae. This technique allowed one to determine widths as small as 50 nm. Data obtained at a flow of 5 nl/s showed that the mixing process was completed at best after 10 µs. The experiments consumed extremely small sample volumes, needing only 25 nl of fluorescein solution and 1 µl of the buffer solution per measurement.

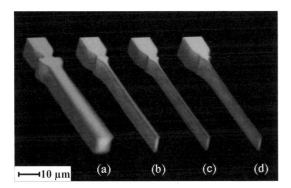

Fig. 3-38. Visualization of hydrodynamic focusing of a fluid layer by means of fluorescence imaging [19].

3.10 Externally Forced Mass Transport, e.g. by Stirring, Ultrasonic Wave, Electrical and Thermal Energy

3.10.1 Dynamic Micromixer Using Magnetic Beads

McCaskill and Schmidt at the Institut für Molekulare Biotechnologie in Jena (Germany) developed an integrated microdevice containing a mixer for biochemical experiments [10]. For amplification reactions the handling of very small volumes (<10 µl) of DNA and RNA under non-batch conditions was required. Thus, the key issue for design of the mixer was to minimize the volume of the components of the microdevice, thereby, among other things, stimulating the development of a micromixer with reduced dead volume.

A dynamic mixing principle, based on stirring of externally driven small particles in a small mixing chamber, was chosen. Magnetic beads made of nickel were enclosed in this chamber. The externally stimulated bead motion resulted in a rapid homogenization of the solutions to be mixed.

3.11 References

[1] Jakubith, M.; *Grundoperationen und chemische Verfahrenstechnik,* Wiley-VCH, Weinheim (1998).
[2] Branebjerg, J., Larsen, U. D., Blankenstein, G.; *"Fast mixing by parallel multilayer lamination"*, in Widmer, E., Verpoorte, E., Banard, S. (Eds.) *Proceedings of the 2nd International Symposium on Miniaturized Total Analysis Systems, µTAS96 – Special Issue of Analytical Methods & Instrumentation AMI,* pp. 228–230, Basel, (1996).
[3] Wilke, H. P., Buhse, R., Groß, K.; *Mischer – Verfahrenstechnische Grundlagen und apparative Anwendungen,* 1st ed; Vulkan Verlag, (1991).
[4] Tatterson, G. B.; *Scaleup and Design of Industrial Mixing Processes,* McGraw-Hill, New York (1994).
[5] Dahlström, D. A.; *"18 Liquid-solid operation and equipment – Phase contacting and liquid-solid processing: agitation of low-viscosity particle suspensions"*, in Perry, R. H., Green, D. W. (Eds.) *Perry's Chemical Engineers' Handbook,* pp. 8–18, McGraw-Hill, New York, (1997).

[6] Tauscher, W., Schneider, G.; "*Statische Mischtechnik*", Chemie Technik **23,** pp. 38–41, 12 (1994).

[7] Mensinger, H., Richter, T., Hessel, V., Döpper, J., Ehrfeld, W.; "*Microreactor with integrated static mixer and analysis system*", in van den Berg, A., Bergfeld, P. (Eds.) *Micro Total Analysis System,* pp. 237–243, Kluwer Academic Publishers, Dordrecht, (1995).

[8] Hessel, V., Ehrfeld, W., Möbius, H., Richter, T., Russow, K.; "*Potentials and realization of micro reactors*", in Proceedings of the "International Symposium on Microsystems, Intelligent Materials and Robots", 27–29 Sept., 1995; pp. 45–48; Sendai, Japan.

[9] Branebjerg, J., Gravesen, P., Krog, J. P., Nielsen, C. R.; "*Fast mixing by lamination*", in Proceedings of the "IEEE-MEMS '96", 12–15 Febr., 1996; pp. 441–446; San Diego, CA.

[10] Schmidt, K., Ehricht, R., Ellinger, T., McCaskill, J. S.; "*A microflow reactor with components for mixing, separation and detection for biochemical experiments*", in Ehrfeld, W., Rinard, I. H., Wegeng, R. S. (Eds.) *Process Miniaturization: 2nd International Conference on Microreaction Technology; Topical Conference Preprints,* pp. 125–126, AIChE, New Orleans, USA, (1998).

[11] Bökenkamp, D., Desai, A., Yang, X., Tai, Y.-C., Marzluff, E. M., Mayo, S. L.; "*Microfabricated silicon mixers for submillisecond quench flow analysis*", Anal. Chem. **70,** pp. 232–236 (1998).

[12] Löwe, H., Ehrfeld, W., Hessel, V., Richter, T., Schiewe, J.; "*Micromixing technology*", in Proceedings of the "4th International Conference on Microreaction Technology, IMRET 4", pp. 31–47; 5–9 March, 2000; Atlanta, USA.

[13] Penth, P.; "*private communication*", (1999) .

[14] Moser, W. R.; Advanced Catalysts and Nanostructured Materials, p. 285 (1996).

[15] Miyake, R., Lammerink, T. S. J., Elwenspoek, M., Fluitman, J. H. J.; "*Micromixer with fast diffusion*", in Proceedings of the "IEEE-MEMS '93", Febr., 1993; pp. 248–253; Fort Lauderdale, USA.

[16] Ehrfeld, W., Golbig, K., Hessel, V., Löwe, H., Richter, T.; "*Characterization of mixing in micromixers by a test reaction: single mixing units and mixer arrays*", Ind. Eng. Chem. Res. **38,** pp. 1075–1082, 3 (1999).

[17] Schubert, K., Bier, W., Brandner, J., Fichtner, M., Franz, C., Linder, G.; "*Realization and testing of micro-structure reactors, micro heat exchangers and micromixers for industrial applications in chemical engineering*", in Ehrfeld, W., Rinard, I. H., Wegeng, R. S. (Eds.) *Process Miniaturization: 2nd International Conference on Microreaction Technology, IMRET 2; Topical Conference Preprints,* pp. 88–95, AIChE, New Orleans, USA, (1998).

[18] Hessel, V., Ehrfeld, W., Haverkamp, V., Löwe, H., Schiewe, J.; "*Generation of Dispersions Using Multilamination of Fluid Layers in Micromixers*", in Müller, R. H., Böhm, B. (Eds.) *Dispersion Techniques for Laboratory and Industrial Production,* Wissenschaftliche Verlagsgesellschaft, Stuttgart, (1999); in press.

[19] Knight, J. B., Vishwanath, A., Brody, J. P., Austin, R. H.; "*Hydrodynamic focussing on a silicon chip: mixing nanoliters in microseconds*", Phys. Rev. Lett. **80,** p. 3863, 17 (1998).

[20] Ehrfeld, W., Gärtner, C., Golbig, K., Hessel, V., Konrad, R., Löwe, H., Richter, T., Schulz, C.; "*Fabrication of components and systems for chemical and biological microreactors*", in Ehrfeld, W. (Ed.) *Microreaction Technology, Proc. of the 1st Int. Conf. on Microreaction Technology,* pp. 72–90, Springer-Verlag, Berlin, (1997).

[21] Hessel, V., Ehrfeld, W., Haverkamp, V., Löwe, H., Schiewe, H.; "*Dispersion Techniques for Laboratory and Industrial Production*", in Müller, R. H. (Ed.) Apothekerverlag, Stuttgart, (1999); in press.

[22] Haverkamp, V., Ehrfeld, W., Gebauer, K., Hessel, V., Löwe, H., Richter, T., Wille, C.; "*The potential of micromixers for contacting of disperse liquid phases*", Fresenius J. Anal. Chem. **364,** pp. 617–624 (1999).

[23] Hessel, V., Ehrfeld, W., Golbig, K., Haverkamp, V., Löwe, H., Richter, T.; "*Gas/liquid dispersion processes in micromixers: the hexagon flow*", in Ehrfeld, W., Rinard, I. H., Wegeng, R. S. (Eds.) *Process Miniaturization: 2nd International Conference on Microreaction Technology, IMRET 2; Topical Conference Preprints,* pp. 259–266, AIChE, New Orleans, USA, (1998).

[24] Suslick, K. S.; in Moser, R. W. (Ed.) *Advanced Catalysts and Nanostructured Materials,* p. 196, (1996).

[25] US 4908154, Microfluidics Corp. Intern.

[26] Möbius, H., Ehrfeld, W., Hessel, V., Richter, T.; "*Sensor controlled processes in chemical microreactors*", in Proceedings of the "8th Int. Conf on Solid- State Sensors and Actuators; Transducers '95 – Eurosensors IX", 25–29 June, 1995; pp. 775–778; Stockholm, Sweden.

[27] Arnold, J., Dasbach, U., Ehrfeld, W., Hesch, K., Löwe, H.; "*Combination of excimer laser micromachining and replication processes suited for large scale production (Laser-LIGA)*", Appl. Surf. Sci. **86,** (1995) 251.

[28] Hessel, V., Ehrfeld, W., Freimuth, H., Löwe, H., Richter, T., Stadel, M., Weber, L., Wolf, A.; "*Fabrication and interconnection of ceramic microreaction systems for high temperature applications*", in Ehrfeld, W. (Ed.) *Microreaction Technology, Proceedings of the 1st International Conference on Microreaction Technology; IMRET 1,* pp. 146–157, Springer-Verlag, Berlin, (1997).

[29] Results of IMM, unpublished

[30] Schwesinger, N., Frank, T., Wurmus, H.; *"A modular microfluid system with an integrated micromixer"*, J. Micromech. Microeng. **6**, pp. 99–102 (1996).

[31] Ehrfeld, W., Hessel, V., Kiesewalter, S., Löwe, H., Richter, T., Schiewe, J.; *"Implementation of microreaction technology in process engineering and biochemistry"*, in Ehrfeld, W. (Ed.) *Microreaction Technology: 3rd International Conference on Microreaction Technology, Proceedings of IMRET 3*, pp. 14–35, Springer-Verlag, Berlin, (2000).

[32] Ehrfeld, W., Hessel, V., Löwe, H.; *"Extending the knowledge base in microfabrication towards chemical engineering and fluid dynamic simulation"*, in Proceedings of the "4th International Conference on Microreaction Technology, IMRET 4", pp. 3–22; 5–9 March, 2000; Atlanta, USA.

[33] Ehrfeld, W., Hessel, V., Haverkamp, V.; *"Microreactors"*, *Ullmann's Encyclopedia of Industrial Chemistry*, Wiley-VCH, Weinheim, (1999).

[34] Villermaux, J., Falk, L., Fournier, M.-C., Detrez, C.; *"Use of parallel competing reactions to characterize micromixing efficiency"*, AIChE Sym. Ser. **88,** p. 286, 6 (1991).

[35] Schubert, K., Bier, W., Linder, G., Seidel, G., Menzel, T., Koglin, B., Preisigke, H.-J.; *"Method and device for performing chemical reactions with the aid of microstructure mixing"*, WO 95/30476, (09.05.1994); Bayer Aktiengesellschaft, Forschungszentrum Karlsruhe GmbH.

[36] Schubert, K., Bier, W., Linder, G., Herrmann, E., Koglin, B., Menzel, T.; *"Verfahren zur Herstellung von Dispersionen und zur Durchführung chemischer Reaktionen mit disperser Phase"*, DE 195 41 265 A1, (06.11.1995); Bayer AG, Forschungszentrum Karlsruhe GmbH.

[37] Schubert, K., Bier, W., Herrmann, E., Menzel, T., Linder, G.; *"Statischer Mikrovermischer"*, DE 195 40 292 C1, (28.10.1995); Forschungszentrum Karlsruhe GmbH, Bayer AG.

[38] Zech, T., Hönicke, D., Fichtner, M., Schubert, K.; *"Superior performance of static micromixers"*, in Proceedings of the "4th International Conference on Microreaction Technology, IMRET 4", pp. 390–399; 5–9 March, 2000; Atlanta, USA.

[39] Schubert, K., Bier, W., Linder, G., Seidel, D.; *"Profiled microdiamonds for producing microstructures"*, Ind. diamond rev. **50,** 5 (1990).

[40] Richter, T., Ehrfeld, W., Gebauer, K., Golbig, K., Hessel, V., Löwe, H., Wolf, A.; *"Metallic microreactors: components and integrated systems"*, in Ehrfeld, W., Rinard, I. H., Wegeng, R. S. (Eds.) *Process Miniaturization: 2nd International Conference on Microreaction Technology, IMRET 2; Topical Conference Preprints*, pp. 146–151, AIChE, New Orleans, USA, (1998).

4 Micro Heat Exchangers

Heat exchangers are utilized to transfer heat efficiently from one streaming fluid over a solid boundary to another streaming fluid. Therefore, sufficiently large contact areas and high temperature gradients, i.e. driving forces for the heat flux, are required. However, to characterize a heat exchanger as a technical device, not only the transfer of heat but also the ratio of heat transfer to pressure loss has to be regarded as an important figure of merit.

A particularly favorable ratio of heat transfer to pressure loss was established during the development of the so-called compact heat exchangers some decades ago. A typical feature of those devices was a splitting of the fluid stream into a large number of partial streams with extremely small characteristic dimensions. As a consequence, the partial streams are characterized by correspondingly low Reynolds numbers, i.e. the partial streams show more a viscous than a turbulent behavior.

To date, plate heat exchangers take into account such design considerations and stimulated design concepts of the majority of micro heat exchangers. Miniaturized platelet-type devices benefit from an increase of the temperature gradients as well as of the specific contacting areas, and hence exert an improved heat transfer. Among the class of platelet-based devices, three generic design concepts can basically be followed (see Figure 4-1). The concepts differ in channel geometry relative to the platelet axis.

First, a number of thin platelets with wide and flat microchannels can be assembled into a stack, forming a multitude of fluid channels alternately filled by two fluids. In a second approach, both fluids are guided through many narrow and deep microchannels within one structured plate. A third approach is similar to the first one, but is based on breakthrough channels.

The first and third approaches need a structuring technique capable of horizontal fabrication, while the second concept can only be realized by vertical structuring, i.e. microfabrication with high aspect ratio.

For the first and second concepts, platelet configurations with microchannels on only one side of the platelet or on both sides can be employed. If the latter option is chosen for the first concept, additional unstructured layers are required to achieve fluid tightness. For the second and third concepts, unstructured layers are needed in any case.

The layer-type architecture of micro heat exchangers generally poses high demands on the interconnection techniques needed since contact between the two streaming fluids has

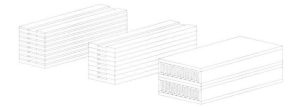

Fig. 4-1. Three generic design concepts for micro heat exchangers and variants to realize them.

to be avoided completely. For this reason, a number of interconnection techniques have been applied, diffusion bonding being so far the method of choice [1–3]. It has to be pointed out that the quality of sealing generally has to be superior to that used for micromixers where contact of the fluids before mixing is, to a certain extent, tolerable.

In the following, the three concepts are introduced in detail.

Micro Heat Exchangers with Wide and Flat Channels
A vast number of systems have been described which were based on alternate stack assembly of two types of platelets to provide a multilayered fluid system of thin layers (see Section 4.1). Most often platelets structured on only one side were used, but two-side structured platelets were realized as well.

In order to achieve a uniform flow distribution, wide and flat channels typically are subdivided by a multitude of microfins which are easily formed by using horizontal fabrication technologies (see Figure 4-1). Based on such a platelet design, micro heat exchangers operating in a counter-, cross- or co-current flow mode have been developed. This choice of flow mode has a major impact on the design of the feed lines and the flow distribution zone, also called header by some authors, while the actual heat exchange region, consisting of a multichannel system, is not affected.

Micro Heat Exchangers with Narrow and Deep Channels
While the first concept has been selected for the majority of micro heat exchanger developments reported, only a few devices of the second and third type have been described so far. One component of an integrated multiphase reactor contains microchannels on both sides of a platelet (see Section 4.2.1) [4, 5]. A device with all channels structured on one side was described as part of an integrated high-temperature gas phase system (see Section 4.2.2) [6].

Micro Heat Exchangers with Breakthrough Channels
At present, only one example of this concept is known (see Section 4.3) [7].

Regarding these three classes, it can be summarized that the conceptual variability of micro heat exchangers currently is not as high as in the case of micromixers. However, many structural variants of devices belonging to one concept have been developed, either as single components or parts of integrated systems. In some cases these devices are commercially available.

Recent results from fluid dynamic simulations may induce an even larger conceptual variety of micro heat exchangers in the near future. The results of the simulations suggest new designs extending the simple platelet architecture concept which was based solely on the improvement of heat transfer due to a decrease of channel dimensions. These new concepts aim at enhancing heat transfer even for a constant channel depth. This is achieved by either shape modification of the channels or material variation of the wall separating the two different fluids.

Three of these new approaches will be presented in Sections 4.4 to 4.6. The first approach is based on blocking structures preventing axial heat losses. The second concept utilizes small cube- or rectangular-like structures for favorable use of entrance flow conditions. Finally, guiding the flow through meander-type structures yields benefits from similar hydrodynamic effects.

4.1 Micro Heat Exchangers with Wide and Flat Channels

4.1.1 Cross-Flow Heat Exchange in Stacked Plate Devices

Construction of Cross-Flow Heat Exchangers
This basic concept of performance improvement through miniaturization has been demonstrated by research work carried out at the Kernforschungszentrum Karlsruhe where, already in the 1980s, developmental work on cross-flow micro heat exchangers was started [3, 8, 9]. Typical examples of such cross-flow micro heat exchangers are shown in Figures 4-2 and 4-3. About 100 platelets with a size of some square centimeters which contain rectangularly shaped microchannels are stacked crosswise and bonded hermetically. As a result, two separated passages for the heat transfer fluid and the process fluid with about 4000 microchannels are formed. The cross-section of a single microchannel shown in Figure 4-2 is 100 μm x 80 μm, and the material thickness between the two fluids in the crossing channels is 20–25 μm. The stack is provided with top and cover plates and connected to suitable fittings for the inlet and withdrawal ducts of the heat exchanger fluids.

The active volume of a micro heat exchanger is typically 1 cm^3 with an inner surface of 300 cm^2 and a heat transfer surface of 150 cm^2. The passages are helium tight with respect both to each other and to the outside. Because of the small dimensions of the channels and the strong connection through bonding, a relatively high operating pressure can be applied. It has been reported that such devices have been tested at pressures up to 25 bar but this should not be the upper limit. The micro heat exchangers have been made from various

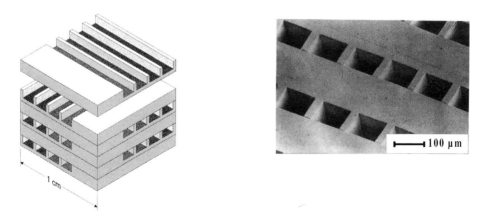

Fig. 4-2. Central component of a cross-flow micro heat exchanger consisting of a stack of crosswise oriented metal platelets [3].

Fig. 4-3. Stack of cross-currently oriented metal foils [12].

materials, e.g. aluminum alloys, copper, silver, titanium and stainless steel. It was demonstrated that the geometrical parameters of the channels, the number of stacked platelets and, of course, the construction of the macroscopic fluid connectors can be adapted flexibly to a wide variety of applications [10, 11].

Microfabrication

Microfabrication of the channels in the platelets was performed by mechanical surface cutting of metal tapes [8]. When aluminum alloys and copper are used, parallel grooves are cut into the tape by means of ground-in monocrystalline diamonds. Since iron alloys are incompatible with diamonds, surface cutting of stainless steel is performed with ceramic microtools which, however, inevitably results in larger tolerances. In the case of titanium the best results are achieved with polycrystalline diamond tools.

After machining the grooves into the tape, cutting into platelets of desired dimensions, which are stacked crosswise to form a cube, is performed. The platelets are connected by

Fig. 4-4. Micro heat exchanger with connections for fluid supply [8].

diffusion bonding to each other, while the cube is connected to the macroscopic fittings by means of electron beam welding (see Figure 4-4).

Characterization of Heat Exchange Performance
In Figure 4-5 the measured thermal power (heat transfer rate) is plotted versus the mass flow rate of water per passage for three micro heat exchangers each having different hydraulic diameters of the microchannels [13]. As to be expected, the thermal power transferred between the fluids increases with decreasing hydraulic diameter of the channels. Of course, a decrease of the hydraulic diameter inevitably results in an increase of pressure loss for a given mass flow and length of the channels, i.e. one has to regard generally the ratio of heat transfer and pressure loss as a fundamental figure of merit.

From the results shown in Figure 4-6 the heat transfer coefficients were calculated (see Figure 4-5) [12]. Even for the maximum flow rate laminar flow conditions should exist in

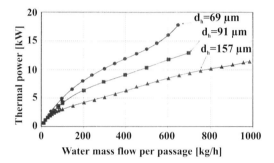

Fig. 4-5. Influence of the hydraulic diameter of the microchannels on the performance of a cross-flow micro heat exchanger, shown by the dependence of heat transfer rate on water mass flow [13].

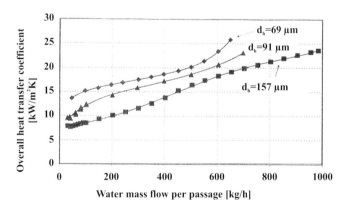

Fig. 4-6. Overall overall heat transfer coefficients versus mass flow for three heat exchangers with microchannels of various hydraulic diameter [12].

each single microchannel. The overall heat transfer coefficients achievable with the micro heat exchanger reach values of more than 20 kW/m^2 K, which considerably exceed those of macroscopic devices.

At a mean logarithmic differential temperature of 59.3 K a power of 19.2 kW was transferred in an active volume of 1 cm^3. The overall heat transfer coefficients were up to 25 kW/m^2 K corresponding to a volumetric heat transfer coefficient of 0.3 kW/cm^3 K. Micro heat exchangers with an active volume of 27 cm^3 and a power up to 200 kW are available [13, 14].

4.1.2 Cross-Flow Heat Exchange Based on Cross-Mixing

Applicability of Classical Heat Transfer Correlations in Cross-Flow Heat Exchangers
The results of a number of studies concerning heat transfer in microchannels were not entirely consistent to the corresponding theoretical descriptions, e.g. regarding the Navier–Stokes equation and classical heat transfer correlations [15–17]. For instance, studies of transfer in micro- and millimeter-sized channels demonstrated higher Nusselt numbers as expected from classical correlations. This holds especially for channels smaller than 1 mm [15, 16]. A number of correlations for Nusselt numbers of flows in these small channels were derived, dependent on the flow regime within the channels.

In order to investigate the validity of these correlations systematically and to exploit this knowledge concerning heat exchanger design practically, researchers at CNRS-ENSIC in Nancy (France) investigated cross-flow compact heat exchangers with millimeter-sized channels (see Figure 4-7) [18]. Two such heat exchangers were made of brass and graphite. A comparison of the overall heat transfer coefficients of both devices was made with classical and non-classical correlations for heat transfer, the latter being derived in micro heat exchangers. Neither of these predictions fitted the experimentally observed behavior of the brass and graphite heat exchangers. Similar to the studies performed in advance, these

Fig. 4-7. Compact heat exchanger with millimeter-sized channels made of copper [18].

results demonstrated a heat transfer improvement for millimeter-sized channels as compared with classical correlations.

A numerical study concerning a single rectangular channel revealed temperature profiles for selected planes. Water was used as a low-conductivity fluid and copper as a highly-conductive channel matrix. It turned out that, due to efficient heat conduction, the temperature profile within the copper material was uniform. In the middle of the channel carrying water as fluid a temperature maximum was observed. Due to comparatively poor thermal conduction of water, the temperature in the central position decreases only slightly along the flow axis, whereas a steep decrease occurs near the channel walls. This limitation due to fluid conduction can in principle be overcome by inducing mass transport. In order to verify this assumption, the simulations were performed on heat transfer assisted by several mixing stages. It could be clearly shown that heat transfer efficiency is notably improved, the higher is the number of mixing stages. Taking transferred power as a measure of efficiency, mixing permits a heat transfer improvement of about 50 %.

Importance of Mixing for Enhancement of Mass Transfer in Minichannels
A comparison of these theoretical predictions with experimental results obtained for the cross-flow heat exchanger was made. Calculated and measured power data were in reasonable agreement. These data did not depend on the material of the heat exchangers when metals were employed. For copper, brass, and zinc devices, simulations predicted equal performance since heat transport in these examples depends only on heat conduction of water. Only in the case of poorly conducting heat exchanger materials such as bakelite is an additional term for heat transfer resistance given. Therefore, overall performance is reduced for such heat exchangers compared with metal-based devices.

Stimulated by the impact of mixing on heat transfer performance, the CNRS researchers designed a heat exchanger with points for cross-mixing (see Figure 4-8). The characteristic feature of this device was two fluid networks consisting of tilted channels of 3 mm diam-

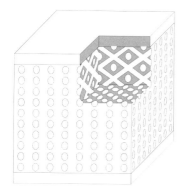
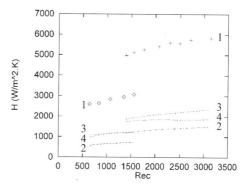

Fig. 4-8. Schematic of a micro heat exchanger with tilted channels which merge at points for cross-mixing (left). Comparison of heat transfer performance based on theoretical (lines) and experimental (dots) data (right) [18].

93

eter which continuously split and merge. This cross-mixing heat exchanger showed a remarkable performance improvement with respect to straight-channel devices made of brass and graphite. Accordingly, the transfer efficiency exceeded the benchmark of the theoretical predictions, achieved by classical or non-classical means, as well.

4.1.3 Counter-Flow Heat Exchange in Stacked Plate Devices

Counter-Flow Microdevices by LIGA Technique
By means of the LIGA process three-dimensional microstructures can be manufactured of nearly any cross-sectional shape and from a wide variety of materials. This flexibility allows the realization of counter-flow micro heat exchangers with an extremely high precision including inlet and withdrawal ducts as well as further auxiliary structures. Compared to other basic configurations of heat exchanging systems counter-flow configurations are the most efficient from a thermodynamic point of view.

The principle of such a micro heat exchanger module is shown in Figure 4-9 [4, 19]. It consists of platelets with an outer frame and an inner thin membrane which are stacked to form a plate-type heat exchanger.

The membranes of the platelets are equipped with parallel fins forming channels for guiding the fluids and increasing the mechanical stability of the device which is of major importance for operating it at high differential pressures. The platelets comprise openings in their corners where two diagonal openings alternately form the inlet and withdrawal ducts for one fluid. The other fluid passes through the remaining openings to the adjacent platelet the design of which, with the exception of a rotary reflection, corresponds to that of the other platelet. On both sides of the membranes which are kept very thin for efficient heat transfer a uniform counter-flow guided by parallel fins is established. The absolute amount of heat transferred in such a stacked device is simply determined by the number of

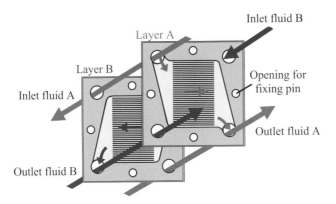

Fig. 4-9. Schematic of design and flow configuration of a plate-type counter-flow micro heat exchanger [4, 19].

Fig. 4-10. Assembled counter-flow micro heat exchanger with PEEK housing and single platelet [4, 19].

platelets which, of course, is limited by the pressure losses in the main inlet and withdrawal channels formed by the openings in the corners of the platelets.

A prototype of a micro heat exchanger based on the principle described above is shown in Figure 4-10. It consists of some stacked platelets clamped between a top and bottom part made of high temperature engineering plastics (PEEK). The top part contains the inlet and outlet ducts for the two fluids of the heat exchanger. The platelets consist of square copper foils with a size of 20 x 20 mm^2 and a thickness ranging from 20 to 500 µm.

The foils are equipped in their center area with 37 nickel fins with a width of 30 µm and a height of some 100 µm forming fluid channels with a width of 300 µm. The heat transfer area amounts to 140 mm^2 per platelet. Each foil is tightly connected with a nickel frame having a thickness corresponding to the height of the fins. The frame comprises also the openings of the fluid ducts and alignment structures.

Microfabrication
The platelets of the micro heat exchanger shown in Figure 4-9 were realized by means of the LIGA process, i.e. deep lithography and subsequent electroforming. In the first step, the complementary pattern of the fins and the frame is generated by means of X-ray lithography in a thick resist layer deposited on an electrically conducting substrate, usually titanium. In the second step, nickel is electrodeposited into the gaps of the lithographic structure forming the fins and the frame. Then, a thin copper film is sputtered on the resist layer as well as on the fins and the frame generated by electrodeposition of nickel. Afterwards, a thick copper layer is grown on a sputtered copper film by means of electrodeposition. The final thickness of the copper foil is adjusted by mechanical aftertreatment. The openings of the ducts are made by drilling.

Since the fabrication method of the platelets used for realization the micro heat exchanger prototype shown in Figure 4-8 is not applicable for mass production, a low-cost replication method was developed. It starts with a platelet of a micro heat exchanger manufactured by deep lithography and electroforming as described above. The precisely electroformed platelet is then used as an electrode to transfer its complementary structure into a hard metal by means of micro electro discharge machining (LIGA-μ-EDM). In this way, a precise tool of high mechanical stability is obtained which allows the low-cost mass production of platelets by means of embossing. Very promising results have been obtained using a relatively soft material like aluminum. In the framework of further development this embossing process will be extended to stainless steel and other materials depending on the chemical properties of the heat exchanging fluids as well as on specific applications of heat exchangers as microreactors and catalyst supports, respectively. Powder injection moulding (PIM), punching and wet etching are further methods which are under development for cost-effective and application-specific mass fabrication of micro heat exchangers or other components of microreactors.

Determination of Heat Exchange Performance
In order to demonstrate the performance of the counter-flow micro heat exchanger some preliminary measurements of heat transfer between a cold (20 °C) and a hot (75 °C) flow of water were performed. For both cold and hot flow, the same flow rates were adjusted and the temperatures were measured by thermocouples located in the center of the inlet and outlet ducts. In the measuring set-up thermal insulation of the stack of heat exchanger platelets was achieved by using polymer material for the top and bottom parts, as well as by inserting the total set-up into a polystyrene-foam box.

Measurements were performed with two different types of platelets which were arranged as stacks of four layers. The first stack consisted of platelets with a 450 μm thick fluid layer, the second set-up was built with thinner platelets having a fluid layer thickness of 100 μm. The measurements of heat transfer rates are plotted versus the volume flow in Figure 4-11.

It turns out that, corresponding to the results obtained with cross-flow micro heat exchangers (see Section 4.1.1), the heat transfer rates increase with decreasing thickness of the fluid layers.

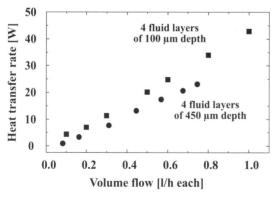

Fig. 4-11. Heat transfer rate versus volume flow for a heat exchanger with four platelets of 100 or 450 μm thickness [20].

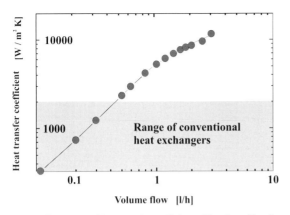

Fig. 4-12. Increase of heat transfer coefficient with volume flow for a counter-flow micro heat exchanger comprising four platelets with a fluid layer thickness of 100 μm. In addition, the performance range of conventional heat exchangers is indicated in the diagram [20].

The dependence of the heat transfer coefficient on the volume flow calculated from the measurements of Figure 4-11 is shown in Figure 4-12.

In correspondence with the results discussed in Section 4.1.1, the values of the heat transfer coefficient achieved in these preliminary measurements considerably exceed those of conventional heat exchangers.

4.1.4 Electrically Heated Stacked Plate Devices

Apart from transferring heat by means of thermal contact of fluids, electrical resistance heating can be used to control the temperature of a microstructured reactor. This approach allows, in particular, the realization of much higher temperatures than achievable using fluids. One example refers to the use of thin meandering stripes, deposited by means of thin film technology, for heating of a microchannel within a silicon gas phase microreactor [21]. The channel wall itself may function as resistive heating element, as evidenced by a catalyst structure for high-temperature reactions consisting of a channel array separated by thin walls [6].

Whilst the heat exchange microstructures were in both examples only part of complex microreaction systems, the use of electrical heaters for micro heat exchanger components solely was described by researchers at the Forschungszentrum Karlsruhe (Germany) [22]. These electrical microchannel heaters should be superior to conventional elements with respect to heating rates, controllability of heating power and compactness.

Ceramic Heaters and Heating Cartridges
For fluid passage, similar microstructures as already described in Section 4.1.1 were employed. These platelets consisted of many parallel microchannels with widths and depths in

Fig. 4-13. Micro heat exchanger composed of microchannel platelet stacks and spacer blocks for insertion of heating cartridges [22].

the range 100 to 200 μm. Two designs of electrical microchannel heaters resulted, depending on the type of heating element, namely ceramic heaters and standard heating cartridges. Both devices were based on the same type of stack architecture, an alternating arrangement of microstructured platelet packs and spacer plates for insertion of heating elements. In the case of the platelet-type ceramic heaters, the spacer plates had rectangular openings, whereas blocks with bore-holes served for insertion of the heating cartridges (see Figure 4-13). Diffusion bonding or electron beam welding were applied for interconnection of all layers of the stack.

The ceramic heaters had a maximum electrical power of 60 W. Since ten heaters were assembled in one stack, a maximum total power of 600 W was achieved for the whole heat exchanger device. The use of heating cartridges resulted in a still higher flexibility regarding power input. Depending on the number and type of heating cartridges, powers of several 100 W up to several kW can be achieved. The device employed in the measurements contained six cartridges capable of 1800 W power generation. Heat transfer coefficients calculated for this device amounted to about 17,000 W/m^2 K, exceeding by far benchmarks from conventional heaters.

Dynamic Response of Ceramic Heaters
The dynamic behavior of the fluid temperature in the microchannels induced by a rectangular voltage pulse of the ceramic heaters was characterized. Using a water flow of 9 l/h, the time delay between onset of electrical voltage and increase of temperature amounted only to 3 s. Within 22 s, the temperature rose from 18 °C to nearly 45 °C. A similar period is needed to cool down after switching off the electric power.

Evaporation Using Heating Cartridges
The performance of the cartridge heated microchannels was tested by an example of use, the evaporation of water. The temperatures of the housing and the fluid at the outlet were

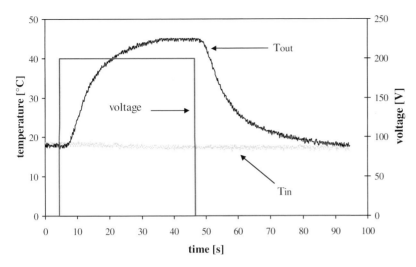

Fig. 4-14. Dynamic behavior of a micro heat exchanger composed of microchannel platelet stacks and spacer plates for ceramic heaters: time-dependent temperature profile after setting a voltage pulse [22].

monitored as a function of the electrical power. Up to a certain set-off point, the corresponding curves of housing and microchannel temperatures were nearly equal. The onset of evaporation resulted in a constant temperature of the water stream of 100 °C, the housing temperature steeply rising.

The ratio of electrical to thermal power, as a measure of heat exchange efficiency, amounts to 93 %. Hence, the use of heating cartridges provides a very efficient means for heat transfer. Another important figure for heat exchanger performance is the heating rate, which amounted to about 300 K/s. In addition to these measurements using water, the microchannel cartridge heater was employed for heating of gases. By these means, mixtures of methane and oxygen, being potentially explosive, were safely heated to 450 °C at a mean residence time of about 6 ms. The corresponding heating rate was calculated to be 72,000 K/s. Using nitrogen as fluid and decreasing the residence time to about 1 ms, even higher heating rates up to 410,000 K/s result.

4.2 Micro Heat Exchangers with Narrow and Deep Channels

4.2.1 Heat Exchanger with One-Sided Structured Channels

A very efficient micro heat exchanger had to be developed by researchers of the Institut für Mikrotechnik Mainz, (Germany) for an integrated high-temperature gas phase microreactor [6]. The heat exchanger was one component of a complex assembled system and was covered by a ceramic disc in which a catalyst structure, i.e. the reaction zone, was embedded.

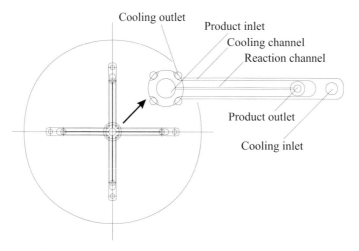

Fig. 4-15. Cross-type micro heat exchanger for co- and countercurrent flow operation, consisting of four product channels surrounded by two cooling channels each [6].

The ceramic material served for thermal insulation of the hot reaction zone from the cold heat exchanger.

After passing the reaction zone, the product gas entered a circular central zone from which four channels were fed, yielding a cross-type design (see Figure 4-15). Each product channel was surrounded by two cooling channels. Finally, the flow was recombined after leaving four outlet ports at the end of the product channels. Due to the parallel arrangement of cooling and product channels, co- or counter-flow operations were possible.

The cross-type channel structure was realized by a two-step μ-EDM process employing die sinking of comb-like electrodes generated by thin-wire erosion. The width of the microchannels was 60 μm, while the depth was about 600 μm, corresponding to an aspect ratio of 10. The microchannels were sealed from each other by means of a laser-cut graphite washer.

Heat transfer calculations indicated that the product gas should be cooled down from about 1000 °C to 120 °C within less than 1 μs. In the framework of the experimental investigations of the Andrussov process (see Section 7.7.1), temperature detection was achieved at a position behind the product channels. Thereby, it could be shown that at least at the oulet ports of the micro heat exchanger the desired temperature of 120 °C was reached.

4.2.2 Heat Exchanger with Double-Sided Structured Channels

Interdigital Heat Exchanger Realized by μ-EDM in Stainless Steel
The key component of an integrated microreaction system for fast liquid/liquid processes are platelets with combined mixer/heat exchanger units [23, 24]. Since it was aimed to

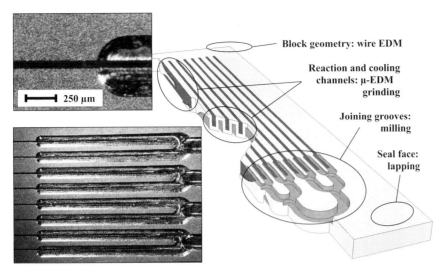

Fig. 4-16. Micro heat exchanger platelet, structured from both sides, for co- and countercurrent flow operation, consisting of eight product channels surrounded by cooling channels with extremely large contact surfaces [23, 24].

build a heat exchange device with large contact areas, a channel configuration with many overlapping channels within one platelet was chosen.

Eight reaction channels were fabricated on the top side of a 1.2 mm thin platelet, while eight cooling channels were introduced from the rear (see Figure 4-16). The two types of channels were arranged side by side yielding an interdigitally overlapping, fluidic structure. In order to achieve efficient heat transfer, 60 µm wide microchannels were chosen. The depth of about 900 µm, corresponding to an aspect ratio of 15, guaranteed a large overlap between the channels as well as a sufficiently high volume flow. The position of the channels had to be realized with a precision of about 3 µm. Due to the parallel orientation of the channels, the fluids were guided in a counter-flow mode.

As the fabrication method, a µ-EDM grinding technique was employed [25, 26]. First, a stress-relieving anneal of the platelets was performed and the outer contour was realized by thin-wire erosion. Afterwards, the microchannels were generated by µ-EDM using a rotating disc electrode. Thereby, extremely small channels of large length and high aspect ratio were accessible. Finally, a lapping process was applied to improve the surface quality.

Temperature measurements within the platelet were performed for a highly exothermic reaction, the synthesis of a vitamin precursor (see Section 6.1). Several thermocouples were set close to the microchannels at different positions, the mixing zone, the outlet zone and in between. At no position could a temperature rise higher than 1 °C be detected, which is considerably different from the technical process showing hot spots as large as 50 °C.

Fig. 4-17. High-aspect ratio micro heat exchanger for co- and countercurrent flow operation, consisting of a thin corrugated layer forming an interdigital channel structure [27].

Interdigital Heat Exchanger Realized by pn-Etch Stop in Silicon
Researchers at the Lund Institute of Technology (Sweden) aimed at realizing silicon micro heat exchangers, consisting of double-sided etched microchannels, with a high aspect ratio and extremely thin channel walls (see Figure 4-17) [27]. Both features should increase the heat exchanger efficiency, by providing large contact surfaces between the two fluids and minimizing the heat transport resistance for conduction through the solid wall separating these media.

Microfabrication of such filigree corrugated structures was achieved by combining anisotropic etching of <110>-oriented silicon, yielding vertical channels, and pn-etch stop technology, as channel walls resulting in thin layers. In the process sequence, one-sided channels were realized by anisotropic etching on a silicon carrier. These channels were phosphorus doped, generating an n-type layer which is resistant against KOH etching. The thickness of this thin layer can be exactly defined by the phosphorus dopant depth. Thereafter, the residual silicon material was completely removed by etching leaving an interdigitated channel structure separated by thin walls. The feasibility of this fabrication concept was evidenced by realizing heat exchanger platelets with microchannels, 360 µm deep and 220 µm wide, separated by a 7 µm thin channel wall.

Several such micro heat exchanger platelets can be stacked on each other to increase throughput. Therefore, the platelets were equipped with specially designed flow entrance and outlet regions.

4.3   Micro Heat Exchangers with Breakthrough Channels

One design concept for micro heat exchangers with breakthrough channels was developed at the Institut für Mikrotechnik Mainz (Germany) [7]. By means of punching, rectangular parts within a platelet were removed yielding a frame-like structure attached to several

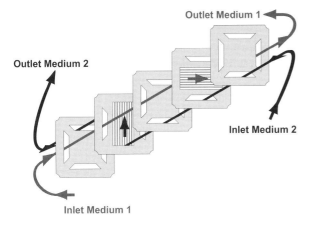

Fig. 4-18. Schematic of assembly of platelets of a micro heat exchanger with breakthrough channels for cross-flow operation [7].

stripes, the interstices of which comprise the breakthrough channels. These channels were covered from the top and bottom by unstructured platelets. Every second frame-like structure has to be oriented at a 90° angle relative to the underlying platelet (see Figure 4-18).

Interconnection of a stack of these platelets results in a cross-flow heat exchanger design (see Figure 4-19). The stack of platelets was interconnected using diffusion bonding. First tests with water, fed at volume flows of 6 l/h and inlet temperatures of warm and cold streams of 78 and 17 °C, respectively, revealed a power transferred up to 220 W.

Fig. 4-19. Photographs of an assembled and disassembled micro heat exchanger with breakthrough channels for cross-flow operation, realized by means of punching in stainless steel [28].

4.4 Axial Heat Conduction

The overall efficiency of heat transfer is related to the transport from one fluid to a channel wall, the conduction through the wall material, and the transfer from the wall surface to a heat exchange fluid. The heat transfer to and from the walls can be increased in miniaturized devices by reduction of linear dimensions. Due to different geometric design parameters, the contribution of heat conduction within the channel wall is significantly different from that of macroscopic devices.

4.4.1 Numerical Calculations of the Influence of Material Choice on Heat Transfer Efficiency

Measurements of heat transfer efficiency of cross-flow micro heat exchangers, manufactured at the Forschungszentrum Karlsruhe (Germany), showed that construction material had a significant influence. Stainless steel microdevices exhibited superior performance over copper devices. This result was explained as being due axial heat conduction, i.e. heat conduction in the direction of fluid flow within the wall material separating the fluid flows. A similar effect is expected to decrease also the performance of counter-current flow heat exchangers by reduction of the temperature gradients, otherwise favorably applied. Therefore, researchers at the Karl-Winnacker Institut of the Dechema, Frankfurt (Germany) performed numerical investigations in order to predict materials of optimized thermal conductivity for the above mentioned purposes.

As a model system, a countercurrent flow microchannel array of shallow geometry was chosen. Initial design values were based on a channel width of 50 μm and a channel depth of 50 μm for both types of fluid channels. The wall separating the two fluids, i.e. representing the plate thickness of real microsystems, amounted to 125 μm, whereas the walls separating the single streams of one fluid were set to 50 μm, i.e. representing the fins. The system was considered to be ideally insulated from the environment. As flow medium, nitrogen was used, entering one group of channels with an inlet temperature of 100 °C, whereas the cooling channel side was set to 20 °C.

Optimization of Heat Transfer Efficiency by Material Choice
As a first result, the heat transfer efficiency of this idealized channel system was calculated as a function of the thermal conductivity in the range 10^{-4} to 10^4 W/m K. It turned out that for materials of very low conductivity the efficiency was low, as to be expected, since radial heat transfer was notably reduced. The corresponding temperature profile in the axial direction shows a discrete temperature level, i.e. no heat exchange takes place.

High-conductivity materials such as copper (394 W/m K) displayed only medium efficiency, amounting to about 50 %. By leveling out of temperature differences in the axial direction, only temperature profiles similar to a co-flow operation are achieved. After a short entrance flow region, both temperature profiles approach a constant temperature half of the difference between the two fluids.

As suggested by the first experiments on heat exchanger performance, materials of reduced conductivity, e.g. glass (about 1 W/m K) gave the best results, reaching efficiencies up to 85 %. The corresponding temperature profile is characterized by a small difference between the temperatures of the two fluids which change nearly linearly with axial flow length.

In addition, two parameter studies were carried out, revealing the influence of wall material and flow rate on optimal heat conductivity for heat exchanger construction material. Increasing wall thickness had only a minor influence due to two counteracting forces. The flow rate, however, has a great impact so that optimum performance is only achieved by using different classes of materials as construction material.

Axial Heat Conduction of Miniaturized Shell-and-Tube Heat Exchangers
Finally, the investigations referred to the question of whether the stacked-plate architecture, which was so far employed for the vast majority of micro heat exchangers, is really the best and most flexible solution regarding heat transfer efficiencies for a broad range of operating conditions. Hence, calculations were performed for a miniaturized analogue of a shell-and-tube heat exchanger. The results demonstrated that by varying the heat exchanger geometry the influence of axial heat conduction can be significantly reduced. Copper- and glass-based micro heat exchangers differ by only about 10 % in performance. Stainless steel devices are nearly equivalent to glass microstructures, whereas for plate-like architectures a decrease of about 15 % was predicted.

Concluding Remarks
The investigations on the impact of axial heat conduction elegantly demonstrate the need for a deeper understanding of the fundamental processes in fluidic microdevices. Without difficulty it can be predicted that similar "suprising" results are waiting to be discovered in the fields of micromixers, microextractors and microchannel catalyst carriers. Obviously, experience of the macro-world sometimes fails when being simply transferred to miniaturized processes. Therefore, it seems that a separate discipline, amalgamating chemical engineering and chemistry with micromechanical knowledge, is ultimately required. The optimization potential of microstructures is definitely large. The establishment of a corresponding technology base, termed microchemical engineering or otherwise, is a future key issue, as important as mass fabrication of microdevices is for sucessful commercial implementation.

4.4.2 The Use of Thermal Blocking Structures

In order to overcome the limitations caused by axial heat conduction, researchers at the Institut für Mikrotechnik Mainz (Germany) investigated a method to decrease the respective contribution for applications where highly conductive materials inevitably have to be used [29]. The theoretical analysis was based on introducing blocking structures, stripes of low conductive materials, into a highly conductive matrix. The conductivity of the blocking structures was chosen to be 1 W/m K.

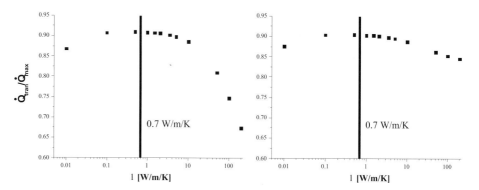

Fig. 4-20. Transferred heat flux as a function of bulk thermal conductivity for channels without (left) and with (right) blocking structures [29].

A normalized parameter, the ratio of transferred heat flux to the thermodynamic limit for this heat flux, was used as a measure of the performance of micro heat exchangers made of different materials with and without blocking structures (see Figure 4-20). The studies clearly demonstrated that a major benefit was only found for materials of a conductivity higher than 100 W/m K, e.g. for silver or copper. Materials of medium conductivity of about 10 W/m K, like the large class of various stainless steels, showed only a small improvement of heat transfer when equipped with blocking structures. This gain in performance does not outbalance the efforts induced by using complex composed platelets for heat exchanger assembly.

To summarize, the studies showed that the use of blocking structures is generally not recommended, but is limited, if feasible at all, to special applications, e.g. those needing silver or copper both as a catalyst and construction material.

4.5 Permanent Generation of Entrance Flow by Fins

When a flow enters a channel or has to be pass an obstacle, an entrance flow region is observed before a fully developed flow profile is established. If this effect is continuously repeated entrance flow conditions can be maintained over the full scale of a device. For instance, checkerboard-like fin structures split a stream continuously in two half which, after development of a full flow profile, have to face a new entrance situation. Thereby, heat and mass transfer are substantially improved. The following discussion will be focused only on the impact on heat transfer.

In order to estimate the influence of using fins, dimensionless heat transfer coefficients, the Nusselt numbers, were compared for microstructures with and without fins (see Figure 4-21). The channel depth was fixed at 250 μm and the size of the fins at 100 μm x 100 μm. A fluid dynamic simulation showed that the Nusselt numbers are at least four times larger

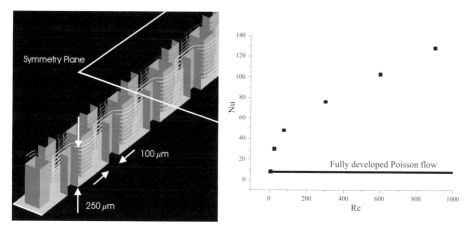

Fig. 4-21. Streamlines and Nusselt numbers in a micro heat exchanger equipped with microfins [29].

with introduction of fins. Moreover, the fins increase the heat exchange surface by a factor of three. Thus, in total the heat transfer in fin-like microstructures is improved by a factor of at least 12.

4.6 Generation of a Periodic Flow Profile by Sine-Wave Microchannels

A similar influence on the flow can be obtained by using sine-wave channels which are structurally similar to a channel system composed of zig-zag-like arranged fins (see Section 4.5) [29]. Due to the curved nature of the sine-wave channels the flow is divided into zones of higher and lower velocity depending on the distance to the channel walls. Thereby, transport of molecules in the radial direction is induced which improves both mass and heat transfer. Again, only the influence on heat transport is reported here. The use of this effect to improve mass transfer, e.g. to accelerate mixing, has already been discussed in Section 3.4.1.

Fluid dynamic simulations were performed for a microchannel of 1 mm width, with an amplitude of 300 µm and a wavelength of 1 mm (see Figure 4-22). The simulations confirmed a similar Nusselt number improvement by use of sine-wave channels as proposed for the use of fins. The heat transfer was actually improved by a factor of 4, at a Reynolds number of 800.

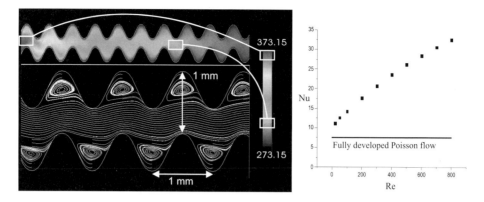

Fig. 4-22. Temperature distribution, streamlines, and Nusselt numbers of a micro heat exchanger with sine-wave microchannels [29].

4.7 Microtechnology-Based Chemical Heat Pumps

Chemical heat pumps for microclimate control have a number of application fields, e.g. for portable cooling systems, car air conditioning, distributed cooling of buildings, and cooling in shipping containers. Researchers at Pacific Northwest National Laboratory (PNNL), Richland (USA) developed a compact microtechnology-based chemical heat pump based on absorber, evaporator, desorber, condenser, heat exchanger, and combustor components (see Figure 4-23) [30, 31]. Due to benefits of mass and heat transfer in microchannel devices, the specific performance of these components could be notably increased, decreasing their size for a given demand.

Fig. 4-23. Components of the microtechnology-based chemical heat pump [30, 31].

Components of the Chemical Heat Pump

Within the absorber and desorber, micromachined contactor units serve for fast gas−liquid transfer between refrigerant vapor and a 50 to 200 μm thin film of a LiBr/water solution. Design targets in membrane contactor development refer to enhancement of permeability and reduction of breakthrough pressure [32]. The heat generated in the absorber is removed by a micro heat exchanger. For absorption of ammonia in water rates of up to 30 W/cm^2 were measured, exceeding conventional absorbers by a factor of ten. The same holds for desorber improvement by means of miniaturization.

The main elements of the evaporator, condenser and heat exchanger are arrays of microchannels, 100 to 300 μm wide and 1 mm deep. In the case of the evaporator, heat transfer coefficients and rates of up to 2 W/cm^{2}K and 100 W/cm^2, respectively, were detected. Compared to technical benchmarks, an improvement by a factor of four is achieved. Similar increases in efficiency were found for the condenser and heat exchanger. The combustor provides a thermal energy of 30 W$_t$/cm^2 at thermal efficiencies of up to 85 %.

An LiBr/water heat pump, consisting of the above mentioned components and relying on water cooling, achieved a cooling power of 350 W at a weight of only 1 kg. Compared to conventional heat pumps, a size reduction of a factor more than 60 can be achieved by using this device. Portable heat pumps using air-cooling were realized as well. For a complete portable cooling system, consisting of the heat pump, heat exchanger, batteries, and fuel, a weight of about 4 to 5 kg results, which is nearly half the value of other existing systems.

The microtechnology-based cooling systems presented by PNNL are relatively complex component assemblies, comprising efficient miniaturized contacting, heat exchange, reaction, and pumping units. All these components have better performance characteristics than existing technical equipment. Nevertheless, the most impressive and unique aspect of the work presented is the interplay between these components, rendering them similar to a "small-scale chemical factory". After a reasonable development time, basic research work regarding component processing was transferred into applications, providing a complete system that meets the requirements defined by application.

4.8 Performance Characterization of Micro Heat Exchangers

The majority of micro heat exchangers have been characterized by determining thermal powers and heat transfer coefficients. These values provide integral information about the efficiency of the corresponding devices. However, regarding reactions with heat release or demand, local measurement of temperature profiles will certainly gain in importance for devices of the next generation. So far, both conventional approaches, e.g. based on thermocouples (see Section 6.2), and more advanced methods, e.g. based on thin-film coated sensors (see Section 7.6.1) were utilized. A further elegant way to reveal temperature profiles is achieved by radiation-based measurements, operating in non-contact mode techniques, as utilized in pyrometers or IR cameras. Especially, IR cameras allow temperature imaging at high spatial resolution and excellent temperature accuracy.

4.8.1 Temperature Profiles of Micro Heat Exchangers Yielded by Thermograms of Infrared Cameras

Researchers at the Fachhochschule Brandenburg (Germany) and the Institut für Mikrotechnik Mainz (Germany) aimed at utilizing a simple, fast, and accurate method for determining temperature profiles of multiple channel systems arranged within one plane [33]. They chose the detection of infrared radiation emission of a body which can be correlated, for known emissivity, to its temperature. Similar thermographic measurements are already state of the art in catalyst screening to measure the activity of several material samples arranged as an array on a silicon chip carrier [34].

IR Camera Investigations via Transparent Windows
Using IR radiation of a few micrometers wavelength, temperatures over a considerable range from minus 50 to about 1500 °C can be monitored at a resolution down to 0.1 K. By using a close-up lens system, a spatial resolution of about 30 µm can be achieved for a measuring field of about 20 mm x 15 mm. In addition, this set-up in principle permits the analysis of fast dynamic processes by real-time thermal measurements at a rate of 1/60 s.

In order to allow measurements in flowing systems, the microreactors had to be equipped with a window which should be transparent for IR radiation. This was exemplarily demonstrated for the falling film microreactor (see Section 8.3), designated for gas/liquid contacting. The corresponding heat exchange system consists of two stacked platelets comprising parallel microchannels which can be operated in countercurrent flow operation.

During experiments, it turned out that the high reflectivity of the stainless steel construction material of the microreactor posed severe problems for temperature detection, in particular when operating close to room temperature. Thus, the whole measuring field had to be shielded from environmental radiation. On this basis, accurate measurements of the temperature profile of the flowing solutions could be gathered.

Heat Exchange Case Study
In a first experiment, the heating process of a flowing 2-propanol solution was monitored. Figure 4-24 shows a temperature profile across the reaction plate of the falling film microreactor consisting of 64 microchannels. It is evident that the "reaction fluid" 2-propanol adapted to the temperature of the heat exchange fluid of about 32 °C. The most remarkable result is that one measurement is sufficient to the compare temperatures of all single microchannels at high resolution. The average difference of the channels' temperatures amounts to less than 0.5 K, for the example studied. Taking into account the precision of these first feasibility measurements could not be set to optimum for several reasons, this value is only an upper-limit estimation of the real performance of the respective micro heat exchanger.

Reaction Case Study
As a model reacting system, the absorption of carbon dioxide in alkaline media inducing heat release during dissolution was investigated. Typical thermograms obtained with and

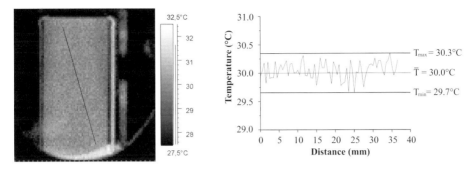

Fig. 4-24. Temperature profile, as determined by an IR camera, across a microstructured platelet comprising 64 microchannels carrying 2-propanol as fluid. This fluid is in thermal contact with a heat exchange fluid flowing in an underlying microstructured platelet [33].

without absorption, which were not corrected for emissivity, are given in Figure 4-25. The channel walls are clearly visible as yellow lines, the fluid in the channels being green to blue colored. However, corrected real-temperature images show a homogeneous temperature distribution across both channels and walls. The temperatures of fluids with and without reaction differ by about 0.5 K, which corresponds to calculations of the temperature rise in case of complete absorption of the gas feed.

Although the experiments discussed above only have scouting character, they outline the high importance of precise determination of analytical data during microreactor operation. This information permits much greater better possibilities for process control of miniaturized devices. However, the application is limited to units equipped with transparent windows. Currently, these are specialized, high-cost devices. Hence, they are ideal tools for information gathering with respect to process optimization. A further application for process control, e.g. for microreactor production units, is not readily attainable to date. The

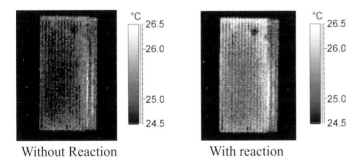

Without Reaction   With reaction

Fig. 4-25. Thermograms of platelets carrying flowing fluids without and with reaction. Platelet of a micro falling film reactor solely wetted by an alkaline solution (left). Platelet of a micro falling film reactor wetted by an alkaline solution exposed to a carbon dioxide gas stream (right) [35].

information gained would, if feasible at all, have to be transferred from measurements of one control unit equipped with a window to a number of other units, all being operated in parallel.

4.9 References

[1] Matson, D. W., Martin, P. M., Steward, D. C., Tonkovich, A. L. Y., White, M., Zilka, J. L., Roberts, G. L.; *"Fabrication of microchannel chemical reactors using a metal lamination process"*, in Ehrfeld, W. (Ed.) *Microreaction Technology: 3rd International Conference on Microreaction Technology, Proceedings of IMRET 3*, pp. 62–71, Springer-Verlag, Berlin, (2000).

[2] Paul, B. K., Peterson, R. B., Wattanutchariya, W.; *"The effect of shape variation in microlamination on the performance of high-aspect- ratio, metal microchannel arrays"*, in Ehrfeld, W. (Ed.) *Microreaction Technology: 3rd International Conference on Microreaction Technology, Proceedings of IMRET 3*, pp. 53–61, Springer-Verlag, Berlin, (2000).

[3] Bier, W., Keller, W., Linder, G., Seidel, D., Schubert, K.; *"Manufacturing and testing of compact micro heat exchangers with high volumetric heat transfer coefficients"*, ASME, DSC-Microstructures, Sensors, and Actuators **19**, pp. 189–197 (1990).

[4] Ehrfeld, W., Gärtner, C., Golbig, K., Hessel, V., Konrad, R., Löwe, H., Richter, T., Schulz, C.; *"Fabrication of components and systems for chemical and biological microreactors"*, in Ehrfeld, W. (Ed.) *Microreaction Technology, Proc. of the 1st Int. Conf. on Microreaction Technology*, pp. 72–90, Springer-Verlag, Berlin, (1997).

[5] Wörz, O., Jäckel, K. P., Richter, T., Wolf, A.; *"Microreactors, a new efficient tool for optimum reactor design"*, in Ehrfeld, W., Rinard, I. H., Wegeng, R. S. (Eds.) *Process Miniaturization: 2nd International Conference on Microreaction Technology, IMRET 2; Topical Conference Preprints*, pp. 183–185, AIChE, New Orleans, USA, (1998).

[6] Hessel, V., Ehrfeld, W., Golbig, K., Hofmann, C., Jungwirth, S., Löwe, H., Richter, T., Storz, M., Wolf, A., Wörz, O., Breysse, J.; *"High temperature HCN generation in an integrated Microreaction system"*, in Ehrfeld, W. (Ed.) *Microreaction Technology: 3rd International Conference on Microreaction Technology, Proceedings of IMRET 3*, pp. 151–164, Springer-Verlag, Berlin, (2000).

[7] Ehrfeld, W., Hessel, V., Kiesewalter, S., Löwe, H., Richter, T., Schiewe, J.; *"Implementation of microreaction technology in process engineering and biochemistry"*, in Ehrfeld, W. (Ed.) *Microreaction Technology: 3rd International Conference on Microreaction Technology, Proceedings of IMRET 3*, pp. 14–35, Springer-Verlag, Berlin, (2000).

[8] Schubert, K., Bier, W., Linder, G., Seidel, D.; *"Profiled microdiamonds for producing microstructures"*, Ind. diamond rev. **50**, 5 (1990).

[9] Bier, W., Keller, W., Linder, G., Seidel, D., Schubert, K., Martin, H.; *"Gas-to-gas heat transfer in micro heat exchangers"*, Chem. Eng. Process. **32**, pp. 33–43, 1 (1993).

[10] Linder, G., Bier, W., Schaller, T., Schubert, K., Seidel, D.; *"Mikrowärmeüberträger und Mikroreaktoren"*, in Proceedings of the "ACHEMA 94, Internationales Treffen für chemische Technik und Biotechnologie; Tagungsband des Symposiums 'Mikrotechnik', 6.–10. Mai, 1994; Frankfurt/M., Germany.

[11] Schaller, T., Bier, W., Linder, G., Schubert, K.; *"Mechanische Mikrotechnik für Abformwerkzeuge und Kleinserien"*, KfK Ber. **5670**, 2. Statuskolloquium des Projekts Mikrosystemtechnik, pp. 45–50 (1995).

[12] Hagendorf, U., Janicke, M., Schüth, F., Schubert, K., Fichtner, M.; *"A Pt/Al$_2$O$_3$ coated microstructured reactor/heat exchanger for the controlled H$_2$/O$_2$-reaction in the explosion regime"*, in Ehrfeld, W., Rinard, I. H., Wegeng, R. S. (Eds.) *Process Miniaturization: 2nd International Conference on Microreaction Technology; Topical Conference Preprints*, pp. 81–87, AIChE, New Orleans, USA, (1998).

[13] Schubert, K., Bier, W., Brandner, J., Fichtner, M., Franz, C., Linder, G.; *"Realization and testing of microstructure reactors, micro heat exchangers and micromixers for industrial applications in chemical engineering"*, in Ehrfeld, W., Rinard, I. H., Wegeng, R. S. (Eds.) *Process Miniaturization: 2nd International Conference on Microreaction Technology, IMRET 2; Topical Conference Preprints*, pp. 88–95, AIChE, New Orleans, USA, (1998).

[14] Schubert, K.; *"Entwicklung von Mikrostrukturapparaten für Anwendungen in der chemischen und thermischen Verfahrenstechnik"*, KfK Ber. **6080**, pp. 53–60 (1998).

[15] Adams, T. M.; *"An experimental investigation on single phase forced convection in microchannels"*, AIChE Symp. Ser. **314,** p. 87 (1997).

[16] Zhang, N., Xin, M. D.; *"Liquid flow and convective heat transfer in microtube"*, in Proceedings of the "China National Heat Transfer", 1991; p. 11189; Beijing.

[17] Peng, X. F., Wang, B. X.; *"Experimental investigation of heat transfer in flat plates with rectangular microchannels"*, Int. J. Heat Mass Transfer **38,** p. 127 (1995).

[18] Luo, L. A., D'Ortana, U., Tondeur, D.; *"Compact heat exchangers"*, in Ehrfeld, W. (Ed.) *Microreaction Technology: 3rd International Conference on Microreaction Technology, Proceedings of IMRET 3*, pp. 556–565, Springer-Verlag, Berlin, (2000).

[19] Ehrfeld, W., Löwe, H., Hessel, V., Richter, T.; *"Anwendungspotentiale für chemische und biologische Mikroreaktoren"*, Chemie IngenieurTechnik **69,** pp. 931–934, 7 (1997).

[20] Ehrfeld, W., Hessel, V., Haverkamp, V.; *"Microreactors"*, Ullmann's Encyclopedia of Industrial Chemistry, Wiley-VCH, Weinheim, (1999).

[21] Jensen, K. F., Hsing, I.-M., Srinivasan, R., Schmidt, M. A., Harold, M. P., Lerou, J. J., Ryley, J. F.; *"Reaction engineering for microreactor systems"*, in Ehrfeld, W. (Ed.) *Microreaction Technology, Proceedings of the 1st International Conference on Microreaction Technology; IMRET 1*, pp. 2–9, Springer-Verlag, Berlin, (1997).

[22] Brandner, J., Fichtner, M., Schubert, K.; *"Electrically heated microstructure heat exchangers and reactors"*, in Ehrfeld, W. (Ed.) *Microreaction Technology: 3rd International Conference on Microreaction Technology, Proceedings of IMRET 3,* pp. 607–616, Springer-Verlag, Berlin, (2000).

[23] Wörz, O., Jäckel, K. P., Richter, T., Wolf, A.; *"Microreactors, new efficient tools for optimum reactor design"*, Microtechnologies and Miniaturization, Tools, Techniques and Novel Applications for the BioPharmaceutical Industry, IBC's 2nd Annual Conference; Microtechnologies and Miniaturization, Frankfurt, Germany, (1998).

[24] Ehrfeld, W., Gärtner, C., Golbig, K., Hessel, V., Konrad, R., Löwe, H., Richter, T., Schulz, C.; *"Fabrication of components and systems for chemical and biological microreactors"*, in Ehrfeld, W. (Ed.) *Microreaction Technology, Proceedings of the 1st International Conference on Microreaction Technology; IMRET 1*, pp. 72–90, Springer-Verlag, Berlin, (1997).

[25] Wolf, A., Ehrfeld, W., Lehr, H., Michel, F., Richter, T., Gruber, H., Wörz, O.; *"Mikroreaktorfertigung mittels Funkenerosion"*, F & M, Feinwerktechnik, Mikrotechnik, Meßtechnik 6, pp. 436–439 (1997).

[26] Richter, T., Ehrfeld, W., Wolf, A., Gruber, H. P., Wörz, O.; *"Fabrication of microreactor components by electro discharge machining"*, in Ehrfeld, W. (Ed.) *Microreaction Technology, Proc. of the 1st International Conference on Microreaction Technology*, pp. 158–168, Springer-Verlag, Berlin, (1997).

[27] Bengtsson, J., Wallman, L., Laurell, T.; *"High aspect ratio silicon micromachined heat exchanger"*, in Ehrfeld, W. (Ed.) *Microreaction Technology: 3rd International Conference on Microreaction Technology, Proceedings of IMRET 3,* pp. 573–577, Springer-Verlag, Berlin, (2000).

[28] Lohf, A., Löwe, H., Hessel, V., Ehrfeld, W.; *"A standardized modular microreactor system"*, in Proceedings of the "4th International Conference on Microreaction Technology, IMRET 4", pp. 441–454; 5–9 March, 2000; Atlanta, USA.

[29] Hardt, S., Ehrfeld, W., vanden Bussche, K. M.; *"Strategies for size reduction of microreactors by heat transfer enhancement effects"*, in Proceedings of the "4th International Conference on Microreaction Technology, IMRET 4", pp. 432–440; 5–9 March, 2000; Atlanta, USA.

[30] Drost, K. M., Friedrich, M., Martin, C., Martin, J., Cameron, R.; *"Recent developments in microtechnology-based chemical heat pumps"*, in Ehrfeld, W. (Ed.) *Microreaction Technology: 3rd International Conference on Microreaction Technology, Proceedings of IMRET 3,* pp. 394–401, Springer-Verlag, Berlin, (2000).

[31] Drost, K., Friedrich, M.; *"A microtechnology-based chemical heat pump for portable and distributed space conditioning applications"*, in Ehrfeld, W., Rinard, I. H., Wegeng, R. S. (Eds.) *Process Miniaturization: 2nd International Conference on Microreaction Technology, IMRET 2; Topical Conference Preprints*, pp. 318–322, AIChE, New Orleans, USA, (1998).

[32] Tegrotenhuis, W. E., Cameron, R. J., Butcher, M. G., Martin, P. M., Wegeng, R. S.; *"Microchannel devices for efficient contacting of liquids in solvent extraction"*, in Ehrfeld, W., Rinard, I. H., Wegeng, R. S. (Eds.) *Process Miniaturization: 2nd International Conference on Microreaction Technology; Topical Conference Preprints*, pp. 329–334, AIChE, New Orleans, USA, (1998).

[33] Hessel, V., Ehrfeld, W., Herweck, T., Haverkamp, V., Löwe, H., Schiewe, J., Wille, C., Kern, T., Lutz, N.; *"Gas/liquid microreactors: hydrodynamics and mass transfer"*, in Proceedings of the "4th International Conference on Microreaction Technology, IMRET 4", pp. 174–186; 5–9 March, 2000; Atlanta, USA.

113

[34] Jandeleit, B., Schaefer, D. J., Powers, T. S., Turner, H. W., Weinberg, W. H.; *"Kombinatorische Material-forschung und Katalyse"*, Angew. Chem. **111,** p. 2649, 17 (1999).

[35] Results of IMM, unpublished.

5 Microseparation Systems and Specific Analytical Modules for Microreactors

Numerous current investigations in microreaction technology deal with miniaturized separation devices using extraction, filtration and diffusion processes. A multitude of publications exist concerning electrophoretic and chromatographic separation devices as well as miniaturized distillation apparatuses. Since these devices are designated as parts of analytical systems for purposes different from chemical reactions, they will not be described in this chapter. The reader is referred to [1–6]. Hence, in the following only microseparation systems specifically developed for chemical processes will be described. In addition, one example of the use of conventional equipment is given which was adapted to the needs of microreactors.

5.1 Microextractors

Extraction processes considered in this section are based on the contact of two immiscible fluids and the resulting solute transfer between the two phases. By means of miniaturization the exchange interface between the two phases is enlarged which, hence, opens the gate to process intensification.

5.1.1 Partially Overlapping Channels

Researchers at the Central Research Laboratories (CRL), Middlesex (U.K.) conducted fundamental work concerning the solute transfer between two immiscible phases in adjacent microchannels, only separated by a stable fluid interface [7]. CRL, in addition, reported on the first applications of these systems and developed a concept for parallelization of a large number of units to enhance volume flow. Later attempts by other researchers tried to stabilize further the contact interface between two channels by the aid of additional microstructures, e.g. using sieve-like structures or conventional porous membranes [8, 9].

Dominant Role of Laminar Flow, Diffusion and Surface Tension in Microchannels
The design concepts developed by CRL were based on the dominance of three effects on the fluid flow and transport properties in microchannels, namely laminar flow, diffusion and surface tension. The theoretical calculations and experimental results aimed at proving this assumption and allowing, thereby, to the reliable prediction of potential applications. As promising fields, liquid/liquid extraction and preparative and analytical chemistry were identified. The design criteria for microsystems were oriented on automated and continuous operation favorably using fast mass transport and eliminating the need of phase separation.

Microfabrication
These criteria are all fulfilled by microdevices with partially overlapping straight channels of a few tens of micrometers width and a few to a few tens of millimeters length (see Figure

Aqueous In   **Organic In**

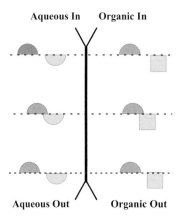

Aqueous Out   **Organic Out**

Fig. 5-1. Diagram of a flow scheme of microcontactors with partially overlapping channels [7]. The scanning electron micrograph of this device is shown in Figure 2-2

5-1). Anodically bonded silicon/glass plates each carrying a single channel were utilized. The nearly rectangular silicon microchannels were realized by sawing, whereas the curved glass microchannels were wet chemically etched. Typical widths and depths of the silicon channels amount to about 50–80 µm and the radius of the semicircular glass channels is set to about 35 µm. The middle axes of both channels were displaced with respect to each other, thereby resulting in partial overlap.

Alignment was performed in a special jig based on a mask aligner stage, capable of channel positioning with ±2 µm precision.

Stabilization of Fluid Interface by Pressure Difference
In the case of flow of immiscible phases in partially overlapping channels, a pressure builds up at the fluid interface, thereby stabilizing a two-phase flow of separate phases and preventing mixing by flow interpenetration. The pressure drop at the interface is a function of surface tension, contact angle of the liquids and distance of liquid layers. The first parameter being constant for a given system, stability can be affected by variation of the latter values, i.e. modification of surface polarity and setting of fluid layer thickness. It turned out that the combination of hydrophilic glass and hydrophobic silicon channels with a fluid layer thickness below 100 µm is sufficient for that purpose.

Further, it could be shown that a decrease in the width of microchannel overlap results in a steep increase of the pressure drop stabilizing the interface. Based on this information, a channel opening of 20 µm was chosen as a comprise between flow stability and mass transfer enhancement.

Diffusion in Parallel Flow of Miscible Phases
The first experiment conducted in the overlapping channels aimed to demonstrate the dominant role of diffusion. For reasons of simplicity, the contacting of miscible fluids was

116

analyzed. A pH induced color change resulted after contacting two aqueous solutions containing hydrochloric acid, sodium hydroxide and a pH indicator. Using flow and plane sheet diffusion equations, effective diffusion coefficients for the acid could be gathered from these measurements. Most of these calculated values were in good agreement with the diffusion coefficient of hydrochloric acid reported in the literature. Being an exception to this behavior, higher effective diffusion coefficients were found when measuring close to the entrance, i.e. the start of liquid contacting. This is due to a lower effective flow rate in this region and, consequently, a longer contact time.

Fractional Fe(III) Ion Transfer in Xylene/Water System
To analyze liquid/liquid contacting, the release and subsequent transfer of formerly complexed Fe(III) ions in a xylene/acidic water system was used. As a measure of extractor efficiency, the concentration of transferred Fe(III) ions divided by the corresponding equilibrium values was taken, referred to as fractional transfer value. Transfer efficiency was defined as fractional transfer multiplied by 100, yielding a percentage value. A large number of microextractor devices with channel lengths of 3, 10 and 30 mm were tested for their transfer efficiency.

A plot of fractional transfer as a function of residence time revealed an increase in transfer efficiency of up to 100 %, i.e. reaching the equilibrium value, after about only 6 s. At a contact time of 2 s, about 50 % transfer occurred. The finding of higher transfer efficiency at the channel entrance region, as demonstrated for miscible fluids (see above), could be confirmed for immiscible phases as well. In a second analysis, the data gathered were plotted as a function of Dt/A, where D denotes the diffusion coefficient, t the organic phase residence time and A the cross-sectional area of the microchannels. Despite notable data scattering, the experimental and theoretical values were in agreement. Therefore, it was concluded that for immiscible phases diffusional transport is the dominant mechanism.

To underline further the accuracy of calculations for the prediction of reactor performance, a comparison of experimental and theoretical values of fractional transfer was performed. Simple plane sheet diffusion equations were applied. With the exception of the

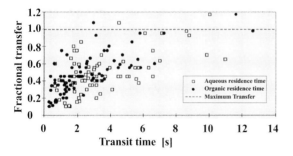

Fig. 5-2. Transfer efficiency of Fe(III) ions in xylene/water as a function of residence time for microcontactors with partially overlapping channels [7].

117

entrance region, a reasonable correpondence of theory and experiment could be confirmed. Moreover, this agreement is not affected by a decrease of the overlapping zone down to 5 μm. Hence, within the practical limits of microstructuring techniques, the interface area provided does not limit mass transport, being dependent only on diffusion, i.e. correlated to fluid layer thickness.

Parallelization Concepts for Large Throughput
In the course of further investigations [10–14], following this first proof of feasibility, CRL produced a series of devices in large numbers, varying in dead space of fluid connectors, channel length, overlap width and channel geometry. In one device, 120 identical contactors were operated in parallel [13]. Again, the ratio of experimental and theoretical fractional transfer was taken as a residence time-independent measure of contactor efficiency. Thereby, it could be shown that, e.g., channel geometry, either being semicircular etched or rectangular sawn channels, has no systematic influence.

Fractional Transfer of Phenol in Octan-1-ol/Water System
As a further model system, the transfer of phenol from an aqueous solution to octan-1-ol was investigated [10]. Within about 1 s the equilibrium concentration was reached. As a particular feature of this process, it turned out that transfer efficiency depends on the overlap width, e.g., resulting in low transfer rates for 5 μm wide openings. This result suggests that interfacial kinetics play an important role in the case of phenol transfer. A theoretical fit of fractional transfer to the normalized diffusion coefficient Dt/A reveals fast effective mass transfer for the channels with large openings (see Figure 5-3). The existence of an initiation period is further evidence for the role of interfacial kinetics in the case of phenol transfer.

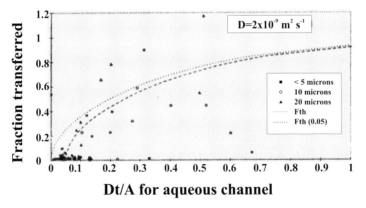

Fig. 5-3. Transfer efficiency of phenol in water/octan-1-ol as a function of residence time for microcontactors with partially overlapping channels [10].

CFD Modeling for Microcontactor Optimization
In a further theoretically oriented contribution, CRL researchers reported on design and process optimization by means of CFD modeling [11, 14]. In addition, the accuracy of CFD based information was compared to that of analytical calculations gained from simple plane sheet expressions. A commercial CFX-4 code was used to solve the non-linear Navier–Stokes equation based on a conservative finite-volume method. This comprises scalar advection–diffusion equations for solute transfer as well as equations for conservation of momentum and mass.

The fractional transfer efficiencies of Fe(III) ions determined by CFD were in perfect agreement with experimental results. By using this method, concentration and velocity profiles were determined even for microchannels of relatively complex cross-section. A comparison of CFD modeling with simple analytical calculations revealed a higher accuracy in the former case. In the light of this comparison, it becomes evident that the scattering of data for plots of transfer efficiency versus Dt/a is probably due to limitations of the simplified analytical approaches. Hence, CFD modeling permits the control of simpler methods, thereby providing an efficient tool with regard to the optimization of contactor geometry and process conditions.

Concluding Remarks
To summarize, the pioneering work on liquid/liquid extraction by the CRL researchers demonstrated elegantly that by using simple, small channel systems considerable improvements of performance can be achieved. The experimental feasibility work was accompanied by detailed simulation studies, facilitating further optimization. Most remarkable is the practical nature of this approach which made the commercialization of devices with hundreds of microchannels possible. Thus, the work of CRL has to be regarded as a milestone within the implementation of microreaction technology in the chemical and pharmaceutical industry.

5.1.2 Wedge-Shaped Flow Contactor

Fundamental Information on Pattern and Stability of Parallel Flows
Researchers at the University of Newcastle and at British Nuclear Fuels, Preston (U.K.), made basic investigations regarding flow patterns of liquid/liquid systems and analyzed the corresponding stability range [15–17]. The first version of microchannel devices applied during these investigations consisted of only one large-sized channel connected to a wedge-shaped flow contactor. The latter microstructure comprised two openings for water feed, embedding a single flow entrance for an organic phase.

Microfabrication
The channels were chemically etched in stainless steel sheets placed between two polished glass plates which were jointly clamped in a stainless steel housing for mechanical sealing. The channel widths were set to 500 to 3000 µm, the channel depths amounted to 100 to

800 µm. As a model system for liquid/liquid contacting, water/kerosene was taken. For visualization a dye was added and propylene glycol was used to vary the viscosity of the aqueous phase.

The design criteria for the channel length and width chosen were oriented on efficient diffusion as theoretically predicted for fluid systems with a Fourier number greater than one [16, 18]. Assuming typical diffusion coefficients of liquid solutes, this correlation demands channel widths ranging from 30 to 100 µm for reasonably short contact times of 1 – 10 s. A further design issue for devices of first generation was to allow simple visualization of flow patterns. Thus, the characteristic channel dimensions were set larger as required for microreaction systems of the next generation.

Stability of Parallel Two-Phase Flow Depending on Operating Conditions
Despite the relatively large fluid channels, a stable parallel liquid/liquid flow could be achieved within a certain range of total volume flow, ratio of flow rates and viscosities (see Figure 5-4). The stability of the flow was influenced in particular by surface forces, whereas other parameters like buoyancy, momentum and viscous stresses were of minor importance. An especially impressive example was the finding of a reversal of phase sequence regarding their density. A stable kerosene flow as lower phase established despite its lower density than water.

Stable parallel flows were obtained when both flow rates were set equal. Decreasing the kerosene flow resulted in a change to a segmented flow pattern characterized by kerosene droplets in a water matrix. The stability of the parallel flow, e.g. as a function of the kerosene flow rate, was increased on decreasing the channel diameter, demonstrating the impact of surface forces. Generally, for wide channels and low flow rates uneven flow profiles were obtained.

Optimized Contactor Geometry Including Flow Splitter
In a second microcontactor device, wedge-shaped structures were used as inlet and outlet elements (see Figure 5-5). Only two instead of three openings, as utilized in the first-generation device, were applied in these structures. The major aim of the investigations using this device was asigned to phase splitting subsequent to establishing a stable two-phase flow. A high-precision separation of the immiscible phases was achieved, accompanied by only minor mixing of the two phases. A quantification of the portion of dispersed phases as

Fig. 5-4. Parallel liquid/liquid flow in a 500 µm x 400 µm channel [15–18].

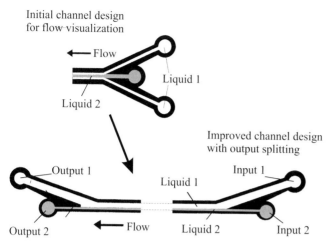

Fig. 5-5. Optimized inlet and outlet design of a microcontactor based on wedge-shaped structures [16, 18].

a function of the kerosene content of the inlet flow revealed a minimum showing nearly zero mixing. These experimental results fitted perfectly data gained by computational calculation.

Reduction of Mass Transfer Limitations for Benzene Nitration
As an example of use, the nitration of benzene with a nitric/sulfuric acid mixture to yield nitrobenzene was carried out. This reaction is sensitive to mass transport, resulting in by-product formation, e.g. benzenes with two or more nitro groups.

$$\text{benzene} \xrightarrow{\text{HNO}_3/\text{H}_2\text{SO}_4} \text{C}_6\text{H}_5\text{NO}_2$$

In order to reduce technical expenditure, a simple set-up was used consisting of two 254 µm wide stainless steel capillaries fed into a 500 mm long stainless steel reactor tube. This set-up serves as a model for the performance of more complex assembled microsystems which were intended for later use. Two reactor tubes were employed with diameters of 254 µm and 127 µm, respectively. Temperature control was achieved by coiling these tubes within a bath equipped with a thermostat. It turned out that much higher conversion was achieved on decreasing the channel diameter and, consequently, the diffusional path. In addition, the good heat transfer properties were evidenced by an enhanced conversion on increasing the bath temperature. At a temperature of 90 °C using the 127 µm tube a conversion of about 49 % was achieved.

Concluding Remarks

It can be concluded that the work yielded information of an original nature concerning the change between parallel and segmented flow and supplementing the CRL results on parallel flow stability. The large channel geometries as well as the relatively inexpensive fabrication process render the technique of parallel liquid/liquid flow to a powerful tool for implementation of microcontactors in chemical production.

5.1.3 Contactor Microchannels Separated by a Micromachined Membrane

Light and Portable Membrane Contactors for Waste Treatment

Researchers at the Pacific Northwest National Laboratory, Richland (U.S.A.), employed microchannel devices separated by a micromachined contactor plate for solvent extraction [8, 9, 19]. Two micromachined channels are separated by a contactor plate, either being a microstructured polyimide foil (Kapton®) or a commercial microporous PTFE foil (Teflon®). This channel configuration allows a co- or counter-current operation of feed and solvent flows, the former containing the solute to be transferred.

Although the field of applications was not specified in detail, the investigations generally aimed at the development of light, compact and efficient devices being portable and allowing transport to remote areas. This includes treatment of contaminated groundwater or nuclear waste, preferably of small processing volumes. In addition, the contactor devices can also be applied for a distributed production at the point of use, e.g. avoiding transport of hazardous chemicals.

Process Engineering Benefits of Membrane Contactors

Apart from specifying detailed applications, benefits in terms of process engineering of miniaturized membrane contactors were clearly identified and contrasted with those of traditional mixer − settler systems. One advantage of film-contacting devices is their higher flexibility with respect to feed − solvent ratios different from one. Moreover, limitations of partitioning equilibrium can be overcome in counter-currently operated film contactors with immobilized interfaces.

Mass transfer in membrane contactors generally is more complex than in other two channel systems with such a barrier, e.g. partially overlapping channels (see Section 5.1.1.) [7]. Resistances can be caused by the flow stream, the contactor plate and the interface. The last two parameters are a function of the membrane porosity and thickness, respectively. By applying theoretical convection − diffusion equations, the theoretical efficiency of the micromachined contactors was determined for comparison with the experimental results.

Micromachining of Microchannel Sheets and Membranes

The membrane contactors were realized in a sheet layer architecture enclosed by two solid endcaps (see Figure 5-6). This concept allows one to combine a large number of shimstock sheets alternately separated by membranes. Feed is performed from header regions distrib-

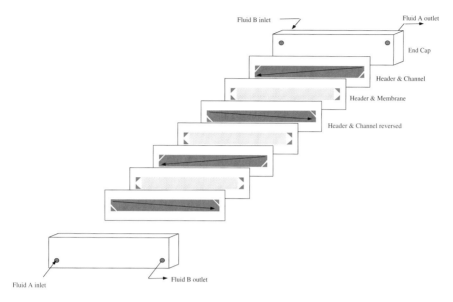

Fig. 5-6. Schematic of assembly of membrane contactors [8, 9, 19].

uting the fluids through holes within the sheets. The sheets are realized in large numbers by photochemical or electrochemical etching methods yielding smooth surfaces.

For sheet interconnection, several options can be followed, either diffusion bonding, mechanical compression or glueing. Only the last two methods are suitable for incorporation of polymeric membranes, while the first method is restricted to membrane materials of high-temperature stability. Diffusion bonding of sheet assemblies was performed by compression under pressure in a jig and subsequent heating under vacuum. This procedure can be applied to materials such as copper, aluminum and stainless steel.

For realization of the membrane, several fabrication strategies were applied. First, an isoporous stainless steel membrane with 130 μm wide pores was produced by means of a photochemical etching process, but not employed during the experiments described below. For that purpose, excimer laser micromachining turned out to be a suitable process. A comparison of a direct writing and mask process revealed that the latter process variant enabled one to achieve much higher porosity of the membrane. By laser drilling using the mask process, an array of isoporous holes of about 25 μm diameter was generated in a 25 μm thin Kapton® foil.

Due to fluid breakthrough, it was necessary to coat this polymeric membrane with PTFE by sputter deposition, adjusting the wetting performance. The porosity of this foil amounted to 26 %. A commercial 3 μm microporous two-layer Teflon® membrane of about 180 μm total thickness was used for comparison. The porosity of the 15 μm thin microporous layer was 44 %. The channel length of 100 mm and width of 10 mm were kept constant, and the depth was set to 200, 300, 400 and 500 μm.

Extraction of Cyclohexanol from Water/Cyclohexane
As a suitable model system, the partitioning of cyclohexanol between water and cyclohexane was chosen. The first data record showed the dependence of extraction efficiency on channel depth. Only for the 400 µm deep channels was a significant mass transfer hindrance observed, while the performance of all other channels was found to be nearly equal. In another set of experiments, the extraction efficiency was compared with the theoretical benchmark derived by the calculations discussed above. Experiments were carried out with a Teflon® membrane and a micromachined plate. The ratio of effluent and feed cyclohexanol concentration was taken as an inverse measure of efficiency and plotted as a function of residence time. The experimental values were compared with theoretical curves for various ratios of effective solute diffusion coefficient in the contactor plate normalized by solute diffusivity in the feed flow (see Figure 5-7).

The experimental data proved the feasibility of the concept by demonstration of an increase of solute in the solvent for both types of membranes with increasing residence time, approaching equilibrium after some minutes of contact. The performance of the microporous membrane was comparable to that of the micromachined membrane, underlining that the difference in porosity is counterbalanced by the variation in membrane thickness.

However, the experiments did not resemble any of the theoretical curves which was tentatively explained by a concentration dependence of the equilibrium partitioning coefficient or flow maldistribution. The latter effect, possibly combined with insufficient wetting of some pores, is also observed when calculating a too high mass transfer hindrance of the contactor plate from the experimental data.

Performance Parameters for Microchannel Contactors
Three parameters were recommended in order to characterize the performance of microchannel contactors, namely the inverse of the Peclet number, residence time, and

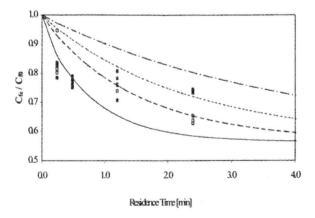

Fig. 5-7. Comparison of theoretical and experimental transfer efficiency of water (cyclohexanol)/cyclohexane as a function of time [8]. Data for the Teflon membrane (light squares), data for the micromachined plate (dark squares), and theoretical data (lines based on various predictions, see [8].

breakthrough pressure, i.e. the maximal pressure drop that the contactor plate can sustain [9]. Based on these parameters, a strategy to size and scale microcontactors was provided. As a design objective for microcontactors, minimization of system size was mentioned, while performing a certain separation task for a given number of theoretical stages and below a breakthrough pressure. As near-term market fields for such reduced-size and light-weight systems, operation during extraterrestrial missions, scrubbing of carbon monoxide in mobile reformer/fuel cell systems, and treatment of radioactive waste were identified.

As a measure for contactor size needed to perform a given separation task, residence time divided by the theoretical number of stages was defined. This figure is analogous to the height equivalent to a theoretical plate (HETP). Both parameters were calculated for conventional sieve tray columns and the microcontactor, based on experimental data for the extraction of acetone from water to 1,1,2-trichloroethane. This comparison revealed that residence time per stage and HETP values can differ by orders of magnitude, thereby indicating the range of size reduction achievable.

These theoretical predictions were experimentally verified by liquid extraction and gas absorption processes. For acetone extraction, equilibrium effluent concentrations were achieved by co-current operation, whereas counter-current flow resulted in separation efficiencies exceeding one theoretical stage. A comparison of these experimental data with the theoretical extraction behavior, gained by numerical solution of transport equations [9], showed a slightly reduced performance, probably due to non-ideal flow behavior. Even more remarkable results in terms of contacting efficiency could be gathered for the carbon dioxide absorption in diethanolamine solutions (see Figure 5-8). For instance, 90 % of the carbon dioxide from a 25 vol. % mixture diluted by nitrogen was absorbed in less than 10s. Similar absorption efficiencies were achieved for carbon monoxide scrubbing from a hy-

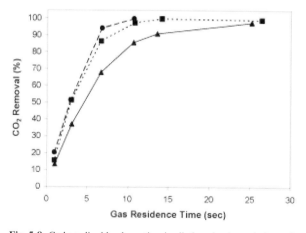

Fig. 5-8. Carbon dioxide absorption in diethanolamine solution using a membrane contactor [9]. Absorption effectiveness as a function of gas residence time for removal of CO_2 from a 25 % CO_2 and 75 % N_2 gas stream into a 10 % diethanolamine solution (triangles), 20 % diethanolamine solution (squares), and 40 % diethanolamine solution (dots) at a liquid residence time of 26 s.

drogen rich stream (70 % N_2, 25 % CO_2, 4 % H_2, 1 % CO) using copper ammonium formate solutions. A reduction to less than 10 ppm was possible within about 1 s.

Concluding Remarks
Similar to the concept of the wedge-shaped flow contactor, the use of membrane microchannel contactors may pave the way for the industrial exploitation of micron-sized transport phenomena. The utilization of various concepts based on simple fabrication techniques certainly increases the range of applicability of these devices. The results with the membrane contactors also indicate that a slightly more complex assembly may result in reduced accuracy of experimental and theoretical results, e.g. as compared with the partially overlapping channels.

Benefits in terms of size and weight reduction in comparison with conventional extraction equipment were clearly outlined. It would be even more instructive to extend this comparison by benchmarking different microcontactors. An evaluation of extraction efficiency normalized by microchannel geometry or total size of the contactor unit would give a deeper insight into the potential of the different concepts.

5.1.4 Contactor Microchannels Separated by Sieve-Like Walls

Other microextraction units use two microchannels separated by sieve-like wall architecture to achieve the separation of two non-dispersed phases [20]. This approach is structurally similar to the concepts using membranes, i.e. sheets with holes, placed between the volumes or channels containing the two phases mentioned (see Section 5.1.4).

Researchers at the Institut für Mikrotechnik Mainz (Germany) realized such an extraction unit based on adjacent channels for the two fluids with slit-shaped openings for solute exchange between two liquid phases. By means of LIGA microfabrication it was possible to realize slit-shaped channels oblique to the flow direction or even slit geometries which may result in a secondary flow in the slit region, improving the exchange process through local convective transport (see Figure 5-9).

Especially in the case of counter-flow operation, large pressure drops can be built up at the interface and, as a result, a bursting of the fluid from one channel into the adjacent channel may occur. Such a bursting can be avoided by proper adjustment of the slit width, i.e. the differential pressure between the fluids can be balanced within a certain limit. Calculations predicted that about three extraction stages can be achieved in such a device with a length of about 160 mm.

5.1.5 Micromixer−Settler Systems

Dynamic Formation of Disperse Phases in Micromixers
By means of microcontactors employing partial contact of two non-disperse phases, mentioned in Sections 5.1.1 to 5.1.5, relatively high specific fluid interfaces are attainable which

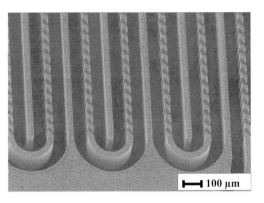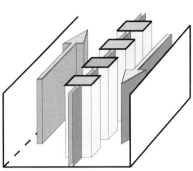

Fig. 5-9. Scanning electron micrograph (left) and schematic representation (right) of an extraction unit with adjacent channels for two fluids with slit-shaped openings, oblique to the flow direction, for exchange between the two phases [20].

are constant over residence time. On the contrary, droplet formation in micromixers exhibits a characteristic time-dependent decrease of this figure (see Section 3.1.3). The maximum values of specific interfaces, reached for a short time of less than 1 s, are a factor of 2 to 5 higher than in case of using microcontactors, but decrease thereafter due to coalescence (see Figure 5-10).

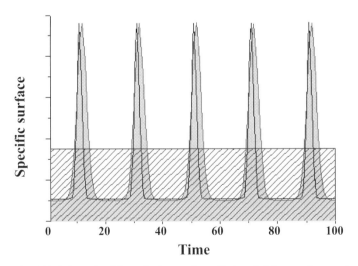

Fig. 5-10. Schematic of dynamic change of specific interfacial areas of a two-phase liquid flow along the flow axis in case of using micromixers. For comparison, a typical constant value of today's microcontactors is depicted [21].

Table 5-1. Liquid model systems for evaluation of extraction efficiency in an interdigital micromixer [21].

Fluid system	Raffinate phase	Extract phase	Extracted component	Surface tension difference [mN/m]	Partition coefficient
1	Water	Toluene	Acetone (10 wt.-%)	33.0	0.82
2	Water	n-Butyl acetate	Acetone (10 wt.-%)	14.0	0.96
3	Water	n-Butanol	Succinic acid (5 wt.-%)	1.7	1.23
4	Water	Methyl-i-butyl-ketone	dl-Panto-lactone (10 wt.-%)	not determined	1.18

In order to test the potential of such a dynamic generation of disperse phases in microfluidic structures, researchers at BASF, Ludwigshafen (Germany) tested interdigital micromixers (see Section 3.1.3), fabricated at the Institut für Mikrotechnik Mainz (Germany), for extraction applications [21]. Special attention was focused on the handling of small volume flows needed for standard mixer−settler systems used in miniplant technology.

Selection of Model Fluid Systems
Four model fluid systems were tested, the first three being recommended by the European Federation of Chemical Engineering [22] and the last system being chosen for internal benchmarking at BASF (see Table 5-1).
The systems differ mainly with respect to their difference in surface tension between aqueous and organic phases. In addition, the diffusion constants of the extracted components vary slightly which is important especially for mass transfer in the settler component, exceeding the micromixer by far in residence time.

Extraction Efficiency
Figure 5-11 shows a typical curve obtained when monitoring phase partition as a function of volume flow, i.e. kinetic energy of the single aqueous and organic streams generated in the interdigital mixing element. Nearly 100 % extraction efficiency could be established for residence times as short as only about 500 ms.
In addition, the flow dependence of the extraction efficiency was analyzed. The shape of the curve is directly related to the interplay between decrease of residence time and increase of specific interface with increasing volume flow. In the range of low volume flows, the increase in specific interface is dominant, i.e. a steep increase in partition values is observed. Above a maximum partition value, at a volume flow of about 800 ml/h, no further significant increase in specific interface is found, which is consistent with previous results determining droplet size [23]. In this range of high volume flows, extraction efficiency decreases slightly due to reduction of residence time.

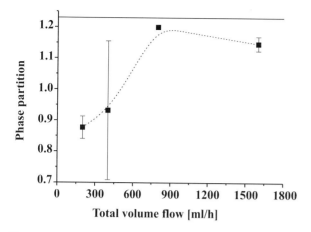

Fig. 5-11. Phase partition as a function of volume flow using an interdigital micromixer for fluid system 3 in Table 5-1, water (succinic acid)/butanol [21].

IMM researchers could confirm this type of dependence of partition values on volume flow by theoretical calculation, taking into account experimental data on droplet size dependence on volume flow for interdigital micromixers and the diffusion of succinic acid (see Figure 5-12).

Moreover, the relative contribution of extraction processes occuring in the micromixer and the settler could be separated by comparative measurements using a settler component

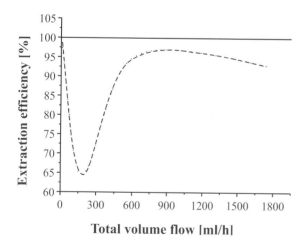

Fig. 5-12. Calculation of phase partition as a function of volume flow for fluid system 3 in Table 5-1, water (succinic acid)/butanol, based on experimental data on droplet size dependence on volume flow for interdigital micromixers [21].

only. Thereby, it could be, for example, shown that in the case of acetone transfer (fluid systems 1 and 2) extraction occurs at least partially within the settler, while for succinic acid and pentolactone transfer (fluid systems 3 and 4) contacting in the micromixer is mostly important.

Series and Parallel Operation of Mixing Elements
In the case of insufficient mass transfer for single contact using one micromixer, e.g. as given for fluid system 4, a series connection of two micromixers yielded much better results. In addition, extraction of similar efficiency was achieved at high volume flows by using micromixer arrays comprising ten parallel operating mixing units. Optimum operation of the mixer array was achieved at a similar pressure loss as for the single micromixer, amounting to an eight-fold increase in volume flow.

Concluding Remarks
The case study on using micromixers performed by BASF is one among a number of impressive examples concerning the flexibility of applying the interdigital mixer concept to different types of reactions and processing. The high extraction partition values obtained despite short contact times clearly prove the high performance attainable in mixing microdevices. For instance, it could be shown that one practical stage corresponds to one theoretical plate at optimum flow. Furthermore, despite decreasing residence time at higher flow rates, enhanced mass transport is observed due to a decrease of the mean droplet size. Therefore, the authors recommend the use of these micromixers as highly efficient tools for miniplant applications.

In this context, it is noteworthy that the current costs of a micromixer, although far from being optimized, are a factor of ten cheaper than the conventional equipment used instead. Moreover, keeping the large volumes and, correspondingly, long residence times of the fluids in the settler in mind, miniaturization of this component seems to be an important issue for future activities.

5.2 Microfilters

Since microreactors inherently suffer from clogging if particle solutions are employed or generated, a filtering step in advance of the reaction may help to prolong the operational life-time of a microreactor. So far, production issues like process reliability or catalyst stability did not play a major role in microreactor research. However, this certainly will be the case the more these systems are implemented in the chemical industry. In particular, irreversibly assembled systems will need careful control of the fluids to be introduced. Instead, reversibly assembled systems can be purified by standard cleaning procedures after disassembly. However, this requires the microreactor operation to be stopped, which is especially detrimental in the case of continuous operation.

At present, filtering of the fluids before entering the current microreactor demonstrators is performed using standard commercial filters. Usually, no details are given in publica-

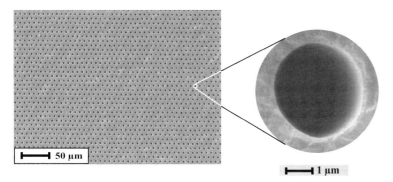

Fig. 5-13. Microfilter with a uniform-sized and -ordered hole array, fabricated by means of UV-lithography and subsequent electroforming [24].

tions, leaving the technical solution for filtering the secret of the particular research group. Regarding future microreactors of enhanced system complexity, the use of large commercial filters may, in some cases, not be sufficient anymore, e.g. if they need to be integrated between two components. This renders the development of microfilters attractive.

5.2.1 Isoporous-Sieve Microfilters

A number of microfilters based on isoporous sieves, i.e. with uniform-sized and -ordered hole arrays, were reported, most often generated by means of lithographic techniques (see Figure 5-13) [24]. These microfilters ideally meet the requirements of complex separation problems, as often found in biochemical separation, where an extreme size exclusion is demanded. Hence the performance of these filters is directly correlated to the precision of the microstructuring technique and superior to existing filter technology, e.g. nuclear tracing filters having a broader pore size distribution.

A distinct disadavantage of microstructured filters for an application in microreaction technology potentially is that they function as so-called "dead-end" filters, i.e. the pores are continuously plugged, thereby decreasing the volume flow at a given pressure drop, until operation has to be completely stopped. This requires a purification or even an exchange of the filter element. A new class of microfilters, termed cross-flow filters, proved to be less sensitive to clogging due to a different mode of filtering operation.

5.2.2 Cross-Flow Microfilters

Cross-flow filters consist of rows of lamellae arranged at a suitable angle of attack relative to the flow direction, resulting in a configuration similar to a Venitian blind (see Figure 5-14) [25]. Due to inertia forces, particles are continuously removed by a concentrated

131

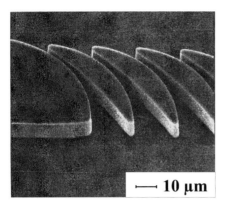

Fig. 5-14. Scanning electron micrograph of a cross-flow filter consisting of lamellae arranged at a suitable angle of attack relative to the flow direction [25].

flow via holes surrounded by the lamellae, while part of the fluid is drawn away through the lamellae, referred to as permeate flow. In any case of clogging of particles on the small slits separating the lamellae, this particle layer can be easily removed by a back-flush procedure, introducing fluid from the permeate region to the feed, i.e. by counter-flow operation. Due to these advantages, cross-flow microfilters are an interesting approach to reduce shear stresses, to increase the operating time of filtration and to purge the filtration unit by reversed flow. Thus, besides applications in chemical processing, they were originally designated for concentrating particles and cells.

The design principle of alternating channels and rows of lamellae is a straightforward approach to an increase in volume throughput by parallelization of many of these units in a planar arrangement. The design is flexible for adaptation to both stack and planar architectures of microreaction systems.

Microfabrication
The microParts company, Dortmund (Germany) realized cross-flow microfilters by deep dry etching of silicon wafers by means of an SF_6/O_2 process. The etching depth was varied in the range 10 to 35 μm, the slits between the lamellae were set to either 2 μm separated by 10 μm lamellae or 4 μm separated by 20 μm lamellae. Apart from using straight lamellae, a third design using curved profiles with decreasing slit width to the feed chamber was realized. The silicon lamellae structures were covered with a glass plate fixed by means of anodic bonding. The pieces were isolated using a wafer saw, mounted in a PMMA block and equipped with capillaries as fluid interconnectors by glueing.

Cross-Flow Filtration Experiments
Using water, a volume flow of about 29 ml/h at a pressure drop of 0.3 bar were achieved. The concentrate to permeate ratio was about 1 to 5 for a fresh filter. In the framework of

experimental investigations, cross-flow filters proved successful for the separation of several types of suspensions including diamond, glass and polymer particles. In the case of highly concentrated suspensions, rapid clogging of the lamellae rows was observed. For a diamond powder, it could be demonstrated that by applying a back-flush cycle, cleaning of the filter structure can be achieved. This was particularly efficient when using cross-flow filters with curved lamellae structures.

5.3 Gas Purification Microsystems

Highly pure gases, especially being free of hydrocarbons, are needed in analytics and emission monitoring. Hydrocarbon-free air is used as a carrier gas in flame ionization detectors. An alternative to the use of expensive gas cylinders containing pure gases is the catalytic combustion, e.g. of air, to convert the hydrocarbons to water and carbon dioxide.

$$CH_4 \xrightarrow[Pt/Al_2O_3]{O_2} CO_2$$

Researchers at the Forschungszentrum Karlsruhe (Germany) realized a microchannel reactor for gas purification which was heated by conventional heating cartridges [26]. The microreactor was based on a platelet stack with 820 microchannels of a hydraulic diameter of 120 µm (see Figure 5-5). The surface of the aluminum platelets was increased by means of anodic oxidation by a factor of 2400 to yield 7.0 m^2/g (see also Section 7.3.2). By nitrogen desorption isotherms, a relatively narrow size distribution of the nanopores generated with an average pore diameter of 41 nm was found. SEM imaging revealed a thickness of the nanoporous aluminum oxide layer of about 20 µm.

Platinum was coated using a wet impregnation procedure yielding loadings between 0.02 and 1.6 wt.-% as evidenced by ICP-emission spectrometry. X-ray fluorescence (WDX) confirmed a homogeneous distribution of the metal over the pore depth. By hydrogen chemisorption, it was found that the average crystallite size amounts to 2.6 nm. The metal dispersion was 43 %.

Methane Conversion
As shown in Figure 5-15, sigmoidal curves were yielded when measuring methane conversion as a function of temperature (120 ppm methane, 20.5 vol.-% O_2 in N_2, 1000 sccm/min, 0.4 wt.-% Pt). With increasing pressure methane conversion was enhanced. At a temperature of 500 °C and a pressure of 5 bar nearly complete conversion, exceeding 99.8 %, was achieved. Increasing methane conversion in the range 100 to 650 ppm resulted in a decrease of methane conversion. This effect is more pronounced at lower temperatures, while at about 520 °C conversion is reduced only by a few percent.

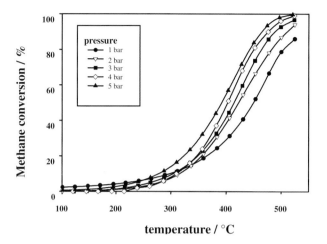

Fig. 5-15. Methane conversion as a function of temperature and pressure [26].

Due to the presence of water vapor as part of the gas mixture in many applications, the dependence of methane conversion on water content was analyzed. Again, only a small decrease in conversion, amounting to a few percent, was found at 520 °C, whereas the greatest impact was detected at low temperatures. As to be expected, metal loading is another important factor influencing methane conversion. Within the range of loadings of 0.02 to 0.4 wt.-% Pt the conversion increases steeply from 4 to 92 %. These studies of catalyst combustion activity were completed by an analysis of catalyst aging during long-term operation for 100 days. No significant reduction in catalyst activity was observed.

Concluding Remarks
The above work conducted successfully demonstrated the realization of a small, compact combustion device as an alternative to large conventional equipment or complicated gas supply systems. Benefits in terms of costs and safety are envisaged. Once more, this study showed the importance of suitable techniques for the generation of nanoporous catalyst carriers and finely dispersed catalyst particles. Thereby, nearly quantitative conversion can be achieved. For commercial implementation, microchannel combustors have to be modified to allow integration into analytical devices.

5.4 Gas Separation Microdevices

A number of gas phase microreactors have been realized and their efficiency concerning a diversity of reactions was tested. However, to utilize these systems as production units in laboratory or industry, strategies for product separation are required. So far, only two pub-

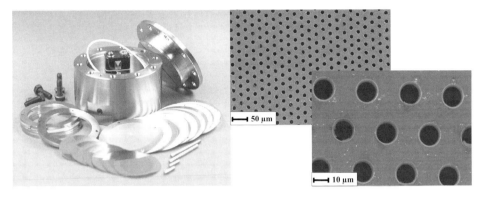

Fig. 5-16. Photograph of disassembled gas separation microdevice (left). Scanning electron micrograph of microstructured support for dense polymeric membrane (right) [27].

lication has focused on a conceptual approach towards gas separation by use of membrane technology [27, 28]. This separation device was specifically developed according to the needs of ethylene oxide synthesis in order to enrich products generated by a microreactor [29]. In a joint action, the group of Wegner at the Max-Planck-Institut für Polymerforschung (Germany) was responsible for the development of suitable polymer membranes, while the Institut für Mikrotechnik Mainz (Germany) carried out fabrication of the gas separation microdevice. The design development of this device was addressed by both groups.

Characteristic features of ethylene oxide separation are the low concentrations of the products in the reactor off-gas. Moreover, unreacted ethylene has to be recycled to be fed into the reactant mixture again. Separation of the contents of gas mixtures having low partial pressures requires the development of efficient tools, e.g. equipped with ultrathin membranes in the range of several tens of nanometers. In order to achieve mechanical stability of such thin films a suitable membrane support has to be designed. Therefore, a microstructured nickel support with an array of isoporous holes was realized by means of a UV-LIGA process (see Figure 5-16). This support was further stabilized by several porous stainless steel sheets. In order to achieve the desired flux, calculations predicted to use a 200 nm thin polymer membrane coated on a support with pore of 60 μm diameter separated by spacings of 80 μm.

The polymer membrane was deposited on the LIGA support by a kind of floatation process using a film spread on water. The membrane-coated support was clamped inside the membrane module, which was designed similarly to conventional flat cells.

By investigation of membrane permeability of various conventional polymers, it was found that due to membrane swelling neither of these materials was suitable for ethylene oxide separation. Hence synthetic activities were started to tailor polymers for the desired application, e.g. to avoid swelling by means of chemical cross-linking.

5.5 Specific Analytical Modules for Microreactors

In a number of applications, analytical techniques had to be adapted to the needs of microreaction systems. Most often off-line analytics were employed. The information about, e.g., using GC, HPLC, MS analytics is spread over the literature describing microreaction systems and their applications. The reader is referred to these original publications and to the corresponding sections within this book.

If measurements are performed on-line or even in-line, more specific or detailed information about the microreactor performance can be gathered, e.g. a three-dimensional profile concerning temperature. In this context, e.g. microsopy investigations of multilaminated layers in micromixers [30–32] and temperature distributions in micro heat exchangers/ reactors [32] can be mentioned. This requires a more specific adaptation of the analytical systems or the applied techniques as in the case of off-line detection. Ultimately, this requires miniaturization of the analytical system itself. For instance, miniaturized temperature sensors were realized by means of thin film techniques [33]. In the following, only those references will be mentioned that rely on the implementation of analytical techniques for microreactors as the central issue, i.e. focusing on methodological developments rather than on reactor microfabrication and testing.

5.5.1 Analytical Modules for In-Line IR Spectroscopy

Researchers at the Forschungszentrum Karlsruhe (Germany) used a movable holder for insertion of a microreactor to perform spatially resolved IR measurements [33, 34]. Positioning of the IR beam could be achieved with micrometer precision and the spot of the IR beam could be reduced to 10 μm. The microreactor was clamped between two IR transparent windows, namely AgCl disks, and inserted in a holder.

The feasibility of this analytical concept was demonstrated for the reaction of silicon tetrafluoride with chlorine to yield various partially and totally fluorinated products [33]. The reaction was carried out in a Y-type micromixer connected to a meandering reaction channel fabricated in a stainless steel plate by μ-EDM techniques. 3D plots of the increase in silicon tetrafluoride and decrease in silicon tetrachloride concentrations were obtained (see Figure 5-17). These plots reveal linear dependences of conversion and yield with axial flow path, i.e. residence time. Sufficient mixing efficiency of the Y-type mixer only after a few micrometers is evidenced by a homogeneous radial distribution of reactant and product concentration close to the mixing inlet.

5.5.2 Analytical Module for Fast Gas Chromatography

Within the framework of a governmentally funded project, a comprehensive approach for the development of a miniaturized gas phase production unit for ethylene oxide synthesis was developed [35, 36]. The activities included the realization of a microreactor (see Sec-

136

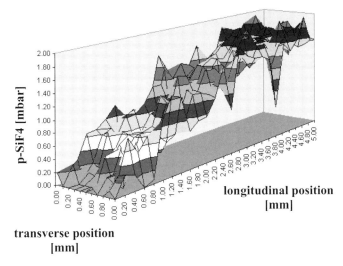

Fig. 5-17. 3D plots of reactant and product concentrations for the fluorination of silicon tetrachloride within a reaction channel [33].

tion 7.5.1), a membrane module for product enrichment (see Section 5.4), and two fast gas chromatographic (GC) analysis modules. Research issues with regard to fast GC analysis were concerned with the development of suitable separation materials for capillaries as well as the realization of fast heating and cooling cycles.

PLOT versus Packed Capillaries
The group of Unger at the Institut für Allgemeine und Angewandte Chemie of the Universität Mainz (Germany) used two different types of capillaries for the separation of typical product mixtures obtained during ethylene oxide synthesis. For efficient separation of permanent gases as well as hydrocarbons and polar compounds, a combination of zeolite and polymeric powder material, respectively, had to be employed. In the case of so-called PLOT capillaries the capillary wall serves as stationary phase. Alternatively, using packed capillary columns, the finely-dispersed separating material is filled into the whole capillary volume. In scouting experiments using laboratory-developed zeolite as well as polymeric materials, it was shown that using packed capillaries is superior to PLOT capillaries in the case of the separation of product mixtures resulting from ethylene oxidation. Thereby, the capillary length could be decreased by a factor of 50 to about 350 to 450 mm and the analysis cycle time was reduced by a factor of two.

Frame GC Module
These short packed capillaries were inserted into a planar module which consisted of an aluminum support comprising channels for insertion of the capillaries and a PEEK frame

137

Fig. 5-18. Photograph of a mounted frame module with integrated heating component for fast GC analysis [36].

carrying heating and cooling components (see Figure 5-18). Cooling in this so-called frame module was achieved by means of a conventional cooling fan as used for personal computers. Heating was accomplished by four ceramic planar heating elements. The temperature was monitored using thermocouples of 0.5 mm diameter inserted into the capillary support. The complete frame module was fabricated by the Institut für Mikrotechnik Mainz (Germany).

In a temperature interval up to 180 °C, a nearly linear heating rate of 2 K/s was determined. As a result only 1 to 1.5 min were needed to achieve the maximum operating temperature. Cooling to room temperature using the fan required about 17 min at a cooling rate of minus 0.5 K/s. Compared with a conventional GC oven, the analysis time needed to separate ethylene oxide from permanent gases and hydrocarbons was reduced from 330 s to about 80 s (see Table 5-2). However, the experiments clearly showed that a further performance improvement of miniaturized GC analysis tools ultimately requires an acceleration of the cooling procedure.

3D GC Module
Therefore, a more complex GC module was developed by the IMM researchers which contained an integrated cooling component in the capillary support (see Figure 5-19). The main components of this module are hollow cylinders fitted into each other. Due to the cylindrical shape of this main part, the whole system was referred to as 3D module. A helical recess in the outer cylinder served as capillary column support. Cooling channels in the inner cylinder were inserted to guide the cooling fluid. The dimensions of the cooling channels were set according to simulation results.

Using the 3D GC module, both heating (4.7 K/s) and cooling rates (minus 5.0 K/s using a 10 bar air flow) could be further improved substantially. The latter value amounts to a 25-fold increase in performance compared with a standard GC oven. All results obtained for the miniaturized frame and 3D modules as well as for the GC oven are summarized in Table 5-2.

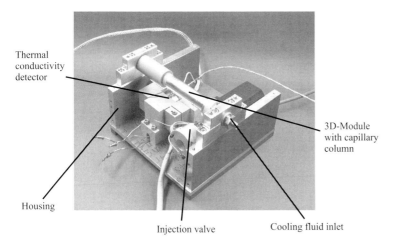

Thermal
conductivity
detector

3D-Module
with capillary
column

Housing

Injection valve

Cooling fluid inlet

Fig. 5-19. Photograph of a mounted 3D module with integrated heating and cooling components for fast GC analysis [35].

Table 5-2. Comparison of heating and cooling performance for two miniaturized modules for fast GC analysis and a standard GC oven.

Heating and cooling equipment	Heating rate [K/s]	Cooling rate [K/s]
GC oven	0.4	–0.2
Frame module	2.0	–0.5
3D module	4.7	–5.0

The favorable heat transfer characteristics of the 3D module result in a further reduction of total analysis time compared with the frame module. Only 55 s are required for separation of the product mixture, and nearly the same time demand is needed for conditioning the capillary prior to the next analysis. Figure 5-20 underlines this notable acceleration of GC cycle time by showing consecutive measurements of five samples performed within 8 min, amounting to a total analysis cycle time of 96 s per sample.

Concluding Remarks
The results obtained for fast GC separation impressively demonstrated further possible improvements with regard to analysis techniques by simple use of the miniaturization concept. Although the single parts were only realized by standard precision engineering techniques, the analysis performance, e.g. in terms of cooling rates, was increased by more than one order of magnitude. Using more precise and advanced microfabrication techniques will certainly lead to a further great leap in technology development.

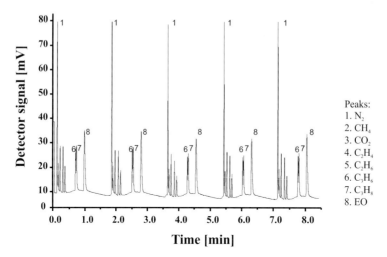

Fig. 5-20. GC plot of a fast multiple analysis obtained by using the 3D module. Five samples are measured representing typical mixtures obtained during ethylene oxide synthesis [35].

It would, moreover, be worthwhile to employ the frame or 3D module, although originally developed for small compact microreactor production units, in the framework of screening applications, e.g. regarding the combinatorial finding of new heterogeneous catalysts for gas phase reactions. So far, most screening applications lack fast, reliable analysis techniques, this being in a number of cases the bottleneck for the whole combinatorial synthesis and screening procedure.

5.6   References

[1]   Woolley, A. T., Mathies, R. A.; *"Ultra-high-speed DNA fragment separations using microfabricated capillary array electrophoresis chip"*, Proc. Natl. Acad. Sci. USA, **91**, pp. 11348–11352 (1994).

[2]   Manz, A.; *"The secret behind electrophoresis microstructure design"*, in Widmer, E., Verpoorte, E., Banard, S. (Eds.) *Proceedings of the 2nd International Symposium on Miniaturized Total Analysis Systems, μTAS96 – Special Issue of Analytical Methods & Instrumentation AMI*, pp. 28–30, Basel, (1996).

[3]   Widmer, H. M.; *"A survey of the trends in analytical chemistry over the last twenty years, emphasizing the development of TAS and μTAS"*, in Widmer, E., Verpoorte, E., Banard, S. (Eds.) *Proceedings of the 2nd International Symposium on Miniaturized Total Analysis Systems, μTAS96 – Special Issue of Analytical Methods & Instrumentation AMI*, pp. 3–8, Basel, (1996).

[4]   Effenhauser, C. S.; *"Integrated chip-based microcolumn separation systems"*, in Manz, A., Becker, H. (Eds.) *Topics in Current Chemistry*, Vol. 194, pp. 51–82, Springer Verlag, Heidelberg, (1998).

[5]   Terry, S. C., Jerman, J. H., Angell, J. B.; IEEE Trans. Electron. Devices **ED-26**, p. 1880 (1979).

[6]   Manz, A., Verpoorte, E., Reymond, D. E., Effenhauser, C. S., Burggraf, N., Widmer, H. M.; *"μ-TAS: miniaturized total chemical analysis systems"*, in Proceedings of the "Micro Total Analysis System Workshop", Nov., 1994; pp. 5–27; Enschede, The Netherlands.

[7]   Robins, I., Shaw, J., Miller, B., Turner, C., Harper, M.; *"Solute transfer by liquid/liquid exchange without mixing in micro-contactor devices"*, in Ehrfeld, W. (Ed.) *Microreaction Technology, Proceedings of the 1st International Conference on Microreaction Technology; IMRET 1*, pp. 35–46, Springer-Verlag, Berlin, (1997).

[8]   TeGrotenhuis, W. E., Cameron, R. J., Butcher, M. G., Martin, P. M., Wegeng, R. S.; *"Microchannel devices for efficient contacting of liquids in solvent extraction"*, in Ehrfeld, W., Rinard, I. H., Wegeng, R. S. (Eds.) *Process Miniaturization: 2nd International Conference on Microreaction Technology; Topical Conference Preprints,* pp. 329–334, AIChE, New Orleans, USA, (1998).

[9]   TeGrotenhuis, W. E., Cameron, R. J., Viswanathan, V. V., Wegeng, R. S.; *"Solvent extraction and gas absorption using microchannel contactors"*, in Ehrfeld, W. (Ed.) *Microreaction Technology: 3rd International Conference on Microreaction Technology, Proceedings of IMRET 3,* pp. 541–549, Springer-Verlag, Berlin, (2000).

[10] Shaw, J., Turner, C., Miller, B., Robins, I., Kingston, I., Harper, M.; *"Characterisation of micro-contactors for solute transfer between immiscible liquids and development of arrays for high throughput"*, in Ehrfeld, W., Rinard, I. H., Wegeng, R. S. (Eds.) *Process Miniaturization: 2nd International Conference on Microreaction Technology; Topical Conference Preprints,* pp. 267–271, AIChE, New Orleans, USA, (1998).

[11] Shaw, J., Miller, B., Turner, C., Harper, R., S., G.; *"Mass transfer of species in micro-contactors: CFD modelling and experimental validation"*, in Widmer, H. M., Verpoorte, E., Barnard, S. (Eds.) *2nd International Symposium on Minaturized Total Analysis Systems, μTAS96,* pp. 185–188, Basel, (1996).

[12] Shaw, J., Turner, C., Harper, R.; *"Visualisation, modelling, and optical measurements in micro-engineered systems"*, in Widmer, H. M., Verpoorte, E., Barnard, S. (Eds.) *2nd International Symposium on Minaturized Total Analysis Systems, μTAS96,* pp. 243–246, Basel, (1996).

[13] Shaw, J., Turner, C., Miller, B., Harper, M.; *"Reaction and transport coupling for liquid and liquid/gas micro-reactor systems"*, in Ehrfeld, W., Rinard, I. H., Wegeng, R. S. (Eds.) *Process Miniaturization: 2nd International Conference on Microreaction Technology, IMRET 2; Topical Conference Preprints,* pp. 176–180, AIChE, New Orleans, USA, (1998).

[14] Bibby, I. P., Harper, M., Shaw, J.; *"Design and optimization of micro-fluidic reactors through CFD and analytical modelling"*, in Ehrfeld, W., Rinard, I. H., Wegeng, R. S. (Eds.) *Process Miniaturization: 2nd International Conference on Microreaction Technology, IMRET 2; Topical Conference Preprints,* pp. 335–339, AIChE, New Orleans, USA, (1998).

[15] Burns, J. R., Ramshaw, C., Bull, A. J., Harston, P.; *"Development of a microreactor for chemical production"*, in Ehrfeld, W. (Ed.) *Microreaction Technology, Proceedings of the 1st International Conference on Microreaction Technology; IMRET 1,* pp. 127–133, Springer-Verlag, Berlin, (1997).

[16] Burns, J. R., Ramshaw, C., Harston, P.; *"Development of a microreactor for chemical production"*, in Ehrfeld, W., Rinard, I. H., Wegeng, R. S. (Eds.) *Process Miniaturization: 2nd International Conference on Microreaction Technology, IMRET 2; Topical Conference Preprints,* pp. 39–44, AIChE, New Orleans, USA, (1998).

[17] Burns, J. R., Ramshaw, C.; *"Development of a microreactor for chemical production"*, Trans. Inst. Chem. Eng. **77**, 5/A, pp. 206–211 (1998).

[18] Crank, J.; *The Mathematics of Diffusion,* 2nd ed; Clarendon Press, Oxford (1975).

[19] Martin, P. M., Matson, D. W., Bennett, W. B.; *"Microfabrication methods for microchannel reactors and separation systems"*, in Ehrfeld, W., Rinard, I. H., Wegeng, R. S. (Eds.) *Process Miniaturization: 2nd International Conference on Microreaction Technology; Topical Conference Preprints,* pp. 75–80, AIChE, New Orleans, USA, (1998).

[20] Ehrfeld, W., Gärtner, C., Golbig, K., Hessel, V., Konrad, R., Löwe, H., Richter, T., Schulz, C.; *"Fabrication of components and systems for chemical and biological microreactors"*, in Ehrfeld, W. (Ed.) *Microreaction Technology, Proceedings of the 1st International Conference on Microreaction Technology; IMRET 1,* pp. 72–90, Springer-Verlag, Berlin, (1997).

[21] Benz, K., Regenauer, K.-J., Jäckel, K.-P., Schiewe, J., Ehrfeld, W., Löwe, H., Hessel, V.; *"Utilisation of micromixers for extraction processes"*, Chem. Eng. Technol. submitted for publication (2000) .

[22] Misek, T., Berger, R., Schröter, J.; *Standard test systems for liquid extraction,* 2nd ed; European Federation of Chemical Engineering, (1985).

[23] Hessel, V., Ehrfeld, W., Golbig, K., Haverkamp, V., Löwe, H., Richter, T.; *"Gas/liquid dispersion processes in micromixers: the hexagon flow"*, in Ehrfeld, W., Rinard, I. H., Wegeng, R. S. (Eds.) *Process Miniaturization: 2nd International Conference on Microreaction Technology, IMRET 2; Topical Conference Preprints,* pp. 259–266, AIChE, New Orleans, USA, (1998).

[24] Ehrfeld, W., Gärtner, C., Golbig, K., Hessel, V., Konrad, R., Löwe, H., Richter, T., Schulz, C.; *"Fabrication of components and systems for chemical and biological microreactors"*, in Ehrfeld, W. (Ed.) *Microreaction Technology, Proc. of the 1st Int. Conf. on Microreaction Technology,* pp. 72–90, Springer-Verlag, Berlin, (1997).

141

[25] Kadel, K., Götz, F., Peters, R. P.; *"Cross-flow filter made using deep dry etching"*, in Ehrfeld, W., Rinard, I. H., Wegeng, R. S. (Eds.) *Process Miniaturization: 2nd International Conference on Microreaction Technology; Topical Conference Preprints,* pp. 96–99, AIChE, New Orleans, USA, (1998).

[26] Wunsch, R., Fichtner, M., Schubert, K.; *"A microstructure reactor for gas purification"*, in Ehrfeld, W. (Ed.) *Microreaction Technology: 3rd International Conference on Microreaction Technology, Proceedings of IMRET 3,* pp. 625–635, Springer-Verlag, Berlin, (2000).

[27] Ehrfeld, W., Hessel, V., Kiesewalter, S., Löwe, H., Richter, T., Schiewe, J.; *"Implementation of microreaction technology in process engineering and biochemistry"*, in Ehrfeld, W. (Ed.) *Microreaction Technology: 3rd International Conference on Microreaction Technology, Proceedings of IMRET 3,* pp. 14–35, Springer-Verlag, Berlin, (2000).

[28] Schiewe, B., Vuin, A., Günther, N., Gebauer, K., Richter, T., Wegner, G.; *"Polymer membranes for product enrichment in microreaction technology"*, in Ehrfeld, W. (Ed.) *Microreaction Technology: 3rd International Conference on Microreaction Technology, Proceedings of IMRET 3,* pp. 550–554, Springer-Verlag, Berlin, (2000).

[29] Kestenbaum, H., Lange de Oliveira, A., Schmidt, W., Schüth, F., Gebauer, K., Löwe, H., Richter, T.; *"Synthesis of ethylene oxide in a catalytic microreacton system"*, in Ehrfeld, W. (Ed.) *Microreaction Technology: 3rd International Conference on Microreaction Technology, Proceedings of IMRET 3,* pp. 207–212, Springer-Verlag, Berlin, (2000).

[30] Branebjerg, J., Larsen, U. D., Blankenstein, G.; *"Fast mixing by parallel multilayer lamination"*, in Widmer, E., Verpoorte, E., Banard, S. (Eds.) *Proceedings of the 2nd International Symposium on Miniaturized Total Analysis Systems, μTAS96 – Special Issue of Analytical Methods & Instrumentation AMI,* pp. 228–230, Basel, (1996).

[31] Knight, J. B., Vishwanath, A., Brody, J. P., Austin, R. H.; *"Hydrodynamic focussing on a silicon chip: mixing nanoliters in microseconds"*, Phys. Rev. Lett. **80**, p. 3863, 17 (1998).

[32] Bökenkamp, D., Desai, A., Yang, X., Tai, Y.-C., Marzluff, E. M., Mayo, S. L.; *"Microfabricated silicon mixers for submillisecond quench flow analysis"*, Anal. Chem. **70**, pp. 232–236 (1998).

[33] Guber, E., Bacher, W.; *"Analytical module for in-line IR spectroscopy of chemical reactions in microchannels"*, in Ehrfeld, W. (Ed.) *Microreaction Technology: 3rd International Conference on Microreaction Technology, Proceedings of IMRET 3,* pp. 617–624, Springer-Verlag, Berlin, (2000).

[34] Guber, A. E., Bier, W., Schubert, K.; *"IR-Spectroscopic Studies of the Chemical Reaction in Various Micromixer Designs"*, in Ehrfeld, W., Rinard, I. H., Wegeng, R. S. (Eds.) *Process Miniaturization: 2nd International Conference on Microreaction Technology; Topical Conference Preprints,* pp. 284–289, AIChE, New Orleans, USA, (1998).

[35] Richter, T., Ehrfeld, W., Hang, T., Löwe, H., Schiewe, J., Yan, X. L., Kurganov, A. A., Unger, K.; *"Integrated heating and cooling module for miniaturized fast GC-analysis"*, in Proceedings of the "4th International Conference on Microreaction Technology, IMRET 4", pp. 141–152; 5–9 March, 2000; Atlanta, USA.

[36] Schiewe, J., Ehrfeld, W., Hang, T., Löwe, H., Richter, T., Yan, X. L., Kurganov, A. A., Unger, K. K.; *"Fast response heating module for temperature programmed GC-analysis in micoreaction systems"*, in Ehrfeld, W. (Ed.) *Microreaction Technology: 3rd International Conference on Microreaction Technology, Proceedings of IMRET 3,* pp. 645–653, Springer-Verlag, Berlin, (2000).

142

6   Microsystems for Liquid Phase Reactions

Compared with the large number of gas phase reactions performed in microreactors up to now, a similar extended list for liquid phase reactions cannot be given. Certainly, this is in some cases due to the higher technical expenditure, e.g. larger pressure drops have to be overcome, especially for viscous liquids, to achieve reasonable throughput. Moreover, reaction time scales of liquid processes are generally longer than those of gas phase processes, typically ranging from some tens of seconds to hours. In the case of slow reaction kinetics, sufficient mass and heat transport may be realized in standard reactors without any need for specialized equipment. Such kinetically controlled reactions will not benefit from miniaturization.

However, there is a common misconception, relating to the long reaction times of most liquid organic reactions. For numerous processes, experimental protocols rely on slow heat transfer or are oriented on completion of reaction rather than being dominated by reaction kinetics. For many liquid reactions these kinetics are even not known. For instance, organometallic reactions are carried out with typical time scales of several tens of minutes in the laboratory and several hours as an industrial process. The same processes may be performed within a few seconds, when transferred from a slow batch process to a fast continuous flow reaction. One impressive example of a reduction of residence time from 5 h to 10s is given below (see Section 6.4) [1].

This clearly shows that process development of liquid reactions (see Section 6.2), in some cases even production (see Sections 6.3 and 6.4), can benefit from miniaturization. As a general guideline, the underlying chemistry should be sufficiently fast, a multiplicity of undesired reactions should exist, and transport phenomena should pose a significant hindrance to the reaction, e.g. referring to highly exothermic and multiphase processes.

Another application field, which is only in its infancy, is the use of continuous flow microreactors for combinatorial synthesis for drug screening or advanced material preparation. Although this application still remains a domain of fully automated batch synthesis using micro titer plates, a first concept for a microreactor with parallel operated channels was proposed (see Section 6.5) [2]. In addition, the extension of combinatorial material research from catalysts and inorganic materials towards polymer synthesis and characterization requires similar miniaturized systems [3].

A common disadvantage of flow-through screening systems, when compared with batch systems, is their higher chemical consumption because of the need to fill complete channel systems connected to extended inlet and outlet structures. A compromise between batch and flow-through synthesis provides rapid serial synthesis based on small liquid plugs of different chemical entities separated from each other.

One possibility for separation can be based on the generation of a two-phase system. In the case of homogeneous catalysis, the use of an interdigital micromixer connected to a capillary as a rapid-serial screening tool was demonstrated recently [4]. Organic plugs consisting of different reactants are separated by aqueous plugs comprising the catalysts.

6.1   Types of Liquid Phase Microreactors

At present, only two examples of specialized microreactor equipment for liquid reactions have been given (see Sections 6.2 and 6.6). In the former case, considerable conversion and heat release were already observed after fractions of a second, requiring a unique design of mixer and heat exchanger [5]. All other investigations were performed in components, most often micromixers connected to channels acting as delay loops or heat exchange devices or both. Hence a classification of liquid microreactors cannot be given currently.

Future results will show if there is a higher demand for more specific devices or if modular constructed components embedded in standardized housings may pave the way for the miniaturization of liquid processes. The generally longer time scale of these processes may prefer the latter concept, by tolerating longer fluid connections within a separate multi-component system versus a compact integrated approach.

In the case of multiphase liquid processes the type of contacting element will play a crucial role. A classification could here be oriented on the definition presented in Chapter 5 for liquid/liquid contactors. In addition, it is likely that liquid reactions requiring solid catalysts or photochemical initiation/activation may demand a unique reactor design. Again, the number of examples reported is not sufficient to outline even reactor categories of only tentative character.

6.2   Liquid/Liquid Synthesis of a Vitamin Precursor in a Combined Mixer and Heat Exchanger Device

The complex chemical structures of vitamins generally require multi-step syntheses when using small molecules as starting material. The total yield of the high-value product vitamin depends on the yields of the single reactions. In particular, the yields of the last steps have to be kept high since a more and more expensive reactant needs to be employed.

BASF researchers performed process development for the synthesis of a vitamin precursor as intermediate product in the framework of a multi-step process [5, 6]. In this case, a cyclization reaction is carried out on an industrial scale as a homogeneously catalyzed liquid/liquid reaction consisting of hexane and concentrated sulfuric acid phases (see Figure 6-1). During the reaction, the reactant, which is initially dissolved in hexane, is converted to an intermediate of more polar nature. Consequently, the latter molecule is extracted into the sulfuric acid phase undergoing a further reaction to yield the vitamin precursor. After completion, the reaction is immediately stopped by diluting the acid phase with water.

A major problem of this process is the formation of byproducts through thermally induced side and follow-up reactions, which reduce the yield. This is caused by the high reaction enthalpy released within a few seconds, since the fast reaction proceeds instantly after contacting the phases.

The standard process was originally performed in a semi-batch production reactor with a yield of 70 %. After many years of production, it was replaced by a continuous process

Reaction scheme

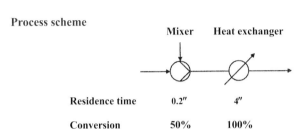

Process scheme

Fig. 6-1. Reaction and process scheme of a homogeneously catalyzed liquid/liquid reaction consisting of hexane and concentrated sulfuric acid phases yielding a vitamin precursor ([7] and by courtesy of Wörz, BASF).

using a mixing pump and heat exchanger. Due to the high reaction rate and knowing the thermal sensitivity of reactant and product, it was tried to keep the residence time in the mixer as short as possible. In spite of a short residence time of only 0.2 s in the mixing pump, a temperature increase of 35 °C occurred.

Since this value amounts to half of the adiabatic temperature rise of the reaction, 50 % conversion occurred already in the mixing pump, i.e. in an uncontrolled thermal environment. The benefits of isothermal processing in the heat exchanger at 50 °C could only be used for completion of the reaction, amounting to a residence time of 4 s. Nevertheless, the yield of 80–85 % of the continuous process exceeded that of the semi-batch process considerably.

From the results obtained on changing their process, the BASF researchers suggested that the improved flow-through reactor still can be optimized further, and aimed in laboratory experiments for reaching this target. However, they failed in the realization of subsecond mixing and fast product separation. The whole procedure took about 2 min, which was far too long and resulted in a dramatic decrease in the yield to 25 %.

At this stage, they decided to choose a microreactor as a laboratory measuring tool knowing about its favorable mass and heat transfer characteristics already from a study concerning a gas phase reaction [5]. The microreactor, designed and manufactured by the Institut für Mikrotechnik Mainz (IMM, Germany), allows the exothermal reaction to be performed under isothermal conditions from the start until completion [8, 9]. A photograph of the microreactor consisting of five stainless steel plates is shown in Figure 6-2. The design specifications were oriented to a limitation of the maximum temperature rise to 1 K and achievement of a total residence time of 4 s, similar to the flow-through process.

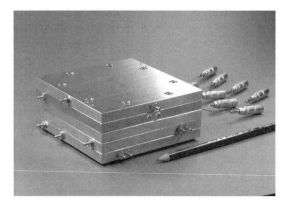

Fig. 6-2. Assembled microreaction system for synthesis of a vitamin precursor comprising integrated static mixers, heat exchangers, reaction channels and delay loop [8, 9].

Process Sequence of the Microreactor

The microreaction system is composed of five plates, namely two feed plates for both reactants, a delay loop plate, a minimixer plate, and a top plate (see Figures 6.3 and 6.4). In the cavities between the feed plates four microstructured stainless steel platelets are inserted which contain a combined contactor and heat exchanger system. This system is composed, apart from a contacting zone, of two groups of eight microchannels each, for reaction and cooling channels, respectively. Thus, the whole microreactor contains 32 reaction channels. In the third plate a delay loop is located which can be used for varying the residence time between 1 and 10 s. In the fourth plate the reaction is quenched by contacting the concentrated sulfuric acid with water in a bifurcation type minimixer. This second reaction is even more exothermic than the main reaction, but the product is no longer thermally sensitive under these conditions.

Process Sequence within the Platelet

The stainless steel platelets integrating contactor and heat exchanger are the central elements of the reactor [8]. The two reactant phases are introduced from opposite sides of the platelet. In an entrance flow region of a few millimeters length the fluids stream in each of the eight microchannels of 60 µm width without being contacted. Thereafter, both channel systems merge, flowing through a breakthrough within the platelet. A sectional view along the flow axis reveals a continuous increase in channel depth of both channel systems, resulting in a tongue-like structure of the platelet material separating them (see Figure 6-5).

After contacting, the two phases are introduced in 60 µm wide and 900 µm deep reaction channels encaved by cooling channels machined into the back side of the platelets. The reaction and cooling channels form a partially overlapping fluidic system. Due to the aspect ratio of 15, large contact surfaces, in addition to small fluidic layers, can be utilized to improve heat transfer. After passing the cooling zone the single streams are recollected in a bifurcation-type structure.

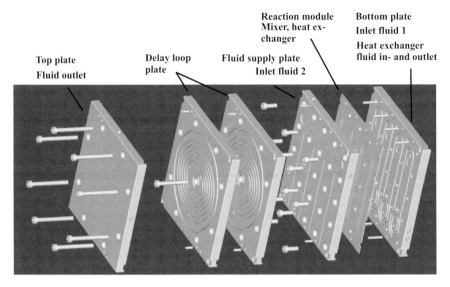

Fig. 6-3. Exploded view of the microreaction system for synthesis of a vitamin precursor comprising two feed plates, integrated static mixers/heat exchanger platelets, delay loop plate, minimixer plate and top plate [10].

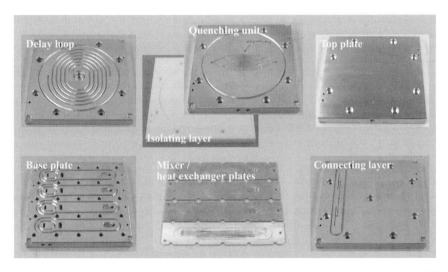

Fig. 6-4. Photograph of the disassembled microreaction system for synthesis of a vitamin precursor [10].

Fig. 6-5. Sectional view of the contacting zone of the reaction platelet.

147

Hydrodynamics of the Two-Phase System

At present, no details about the fluid dynamics of the two-phase system are known, e.g. if two continuous phases are formed, i.e. bilayer, or a disperse system similar to an emulsion results. No information either on the initial formation or on the stability of such generated fluid systems is available.

The BASF reaction is known not to be extremely sensitive to mixing, because mass transfer is accelerated strongly by reaction. For a second liquid/liquid reaction, investigated by another company, mixing quality was more crucial. In this case, fairly good results were obtained as well, indicating that the interfacial areas achieved might be higher than for a simple bilayer system. However, a deeper insight can only be obtained by future investigations relying on a direct visualization of the streams or model reactions for reactor benchmarking.

Microfabrication

Micro electro discharge machining (μ-EDM) processes were employed for the fabrication of the contactor/heat exchanger platelets [8, 11], while the residual parts of the microreactor were realized by conventional precision engineering techniques, in particular by micromilling. For the microchannels with a high aspect ratio, a new micro discharge machining process had to be developed at IMM, termed microerosive grinding. This refers to a micro-die sinking process using a rotating disc electrode of some tens of micrometers thickness, preferably made of tungsten.

Precise positioning of electrodes to the workpiece was required for a highly accurate alignment of reaction and cooling channels. The influence of electrode wear on the channel geometry during the grinding process was eliminated by an automatic compensation procedure. Meanwhile, the grinding technique has been replaced by a standard die sinking process with a fin-shaped electrode manufactured by micro-wire erosion. This modification allows the realization of all channels parallel in one step, whereas grinding can only serially fabricate one channel after the other.

As platelet material, a special stainless steel alloy was chosen being resistant under highly corrosive reaction conditions. It turned out that cleaning of such platelets with high-aspect ratio microchannels is possible, even after complete clogging with polymeric precipitates. The latter was only observed when using process parameters which were far from the standard of the large-scale reactor. In all other cases, neither notable material deposition nor chemical attack of the reactor material was detected.

Experimental Results

During experiments, it was aimed to start with the standard parameters, namely a reaction time of 4 s and a temperature of 50 °C. From this data set, the residence time should be further varied, e.g. to analyze if improvements can be achieved at shorter contact times [5, 7].

The confidence in the microreactor as a laboratory measuring tool was confirmed by the fact that already the first result, referring to a residence time of 4 s, gave a yield of 80–85 % corresponding to that of the technical benchmark (see Figure 6-6). As expected, longer

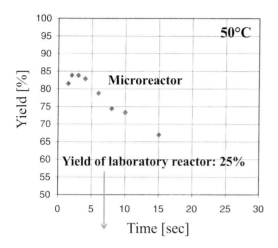

Fig. 6-6. Feasibility of using the microreactor as a laboratory tool for process development, as evidenced by attainment of a yields comparable to the BASF large scale process [7].

residence times led to a strong decrease of the yield, since the labile product is longer than necessary exposed to the hot reaction temperature. For instance, prolongation of the residence time to 15 s decreased the yield to about 65 %. This dependence corresponds to the findings using the conventional laboratory technique which inevitably results in a long residence time (yield 25 %).

A decrease of residence time, however, showed a similar influence on reactor performance. The decrease of yield is most likely due to the kinetics of the reaction, being too slow to complete conversion below 4 s. Thus, it turned out that, by chance, the residence time used for the large-scale reactor is the optimum choice. No further improvements can be expected by tuning the reaction by variation of this process parameter.

As a second parameter for optimization, the variation of the reaction temperature was investigated. Lower process temperatures result in reduced technical expenditure and may increase selectivity due to the thermal sensitivity of reactant and product. In order to adapt to the reaction kinetics, the range of residence times to be investigated had to be extended. Therefore, experiments were run at 20 °C for 1 to 120 s (see Figure 6-7). At a residence time of 30 s, a maximum yield of nearly 95 % was achieved which considerably exceeds that of the continuous technical process. Accordingly, this result shows that due to temperature variation the amount of byproducts can be reduced by more than 50 %.

Introducing Modularity for the Liquid/Liquid Microreactor
In order to extend the operational range of the liquid/liquid microreactor described above, in particular concerning variation of residence time and performance of multi-step syntheses, IMM researchers developed a modular design [10]. Thereby, much greater flexibility regarding the sequence of the components and, hence, the type of processing was achieved.

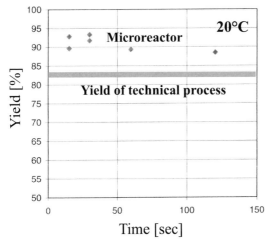

Fig. 6-7. Increase in yield of the vitamin precursor by about 15 % compared with the technical process by finding of optimized process parameters [7].

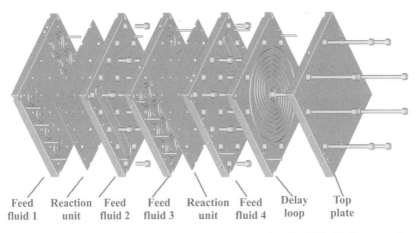

Feed fluid 1   Reaction unit   Feed fluid 2   Feed fluid 3   Reaction unit   Feed fluid 4   Delay loop   Top plate

Fig. 6-8. CAD drawing of one type of assembly of the modular designed liquid/liquid microreactor for performance of two consecutive reactions.

For instance, additional delay loop plates can be inserted and several contactor/heat exchanger units can be arranged in series. Furthermore, two or more reaction can be carried out subsequently (see Figure 6-8).

Concluding Remarks

The measurements of the BASF researchers clearly demonstrated that microreactors provide a clear basis for decisions concerning modifications of large-scale reactors which may cause enormous costs. A similar precise setting of operating parameters, in particular short residence times and isothermal boundary conditions, was not possible by the existing laboratory equipment of BASF. The results were obtained in a short time, e.g. within a few weeks. Due to the reversible connection of the plates of the microreactor, visible control was possible as well as cleaning of the individual parts. This modular assembly allowed, in case of any malfunction, a single part to be replaced instead of exchanging the whole reactor. This, again, saves costs and time, which would otherwise be needed for the manufacturing process and interrupt the experimental investigations.

6.3   Acrylate Polymerization in Micromixers

Radical polymerization processes are employed industrially for the synthesis of a vast number of polymers and copolymers. The production of acrylates, such as poly(methyl-methacrylate) (PMMA) or poly(acrylic acid) (PAA), so far was realized by batch and semibatch processing. Recently, potential benefits of continuous radical polymerization have been identified, such as enhanced process reliability and reactor efficiency as well as increased safety.

Premixing of Solutions fed into a Tube/Static Mixer Reactor
To evaluate this potential, a tube reactor equipped with static mixers was tested in a case study conducted by researchers at Axiva (formerly Aventis), Frankfurt (Germany) [12]. As a crucial feature, uniform local concentration profiles of initiator and monomer had to be established directly after mixing these reactants. This so-called micromixing process was performed by a 5 mm static mixer. The reactant solutions entered this premixing tool before being fed into a tube reactor of 5 to 20 mm diameter and 23 to 60 m length. The residence time was set to 40 min and operating temperature of maximum 150 °C and a pressure drop of maximum 16 bar were applied. A total throughput of 6−8 kg/h was achieved. The premixer was operated in the laminar regime at a Reynolds number of about 400.

Due to imperfect mixing in the static premixer, precipitation of polymers occurred (see Figure 6-9). This led to significant fouling problems for the static mixer during solution polymerization of various acrylates. An analysis by gel permeation chromatography (GPC) reveals that this is due to the formation of an insoluble fraction of high-molecular-weight polymers, ranging from 10^5 to 10^6 g/mol. The formation of the latter is due to local variations of initiator concentrations on a micro-scale, influencing the sequence of chain start, chain growth and termination steps. In particular, a local reduction of the number of termination processes with respect to chain growth leads to an increase of molecular weight.

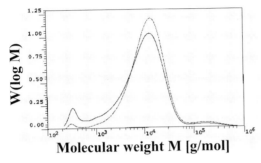

Molecular weight M [g/mol]

Fig. 6-9. Photograph of a macroscopic static mixer clogged due to high-molecular-weight polymer precipitation (top). GPC plot revealing the respective polymer fraction (bottom) [12].

Use of a Micromixer as Premixer

In order to prevent this feed segregation, interdigital micromixers were employed. To achieve a reasonable throughput, an array with ten interdigital mixing units fabricated by IMM was utilized [13]. Each unit consisted of 36 microchannels of a width of 25 μm made of silver. The generation of thin lamellae of initiator and monomer solutions actually sigificantly reduced fouling problems (see Figure 6-10). The corresponding GPC plot shows only the main polymer fraction with a number average close to 10^4 g/mol. Thus, the product quality, in terms of polymer uniformity, was notably improved by using a miniaturized mixing device.

Pre-Basic Design for Numbering-Up

In order to transfer the results obtained on a lab-scale with the micromixer to a production process, the classical route of up-scaling the device cannot be followed. The performance of micromixers is directly correlated to their characteristic channel dimensions. Hence, only numbering-up concepts based on parallel operation of identical devices or units, as already proposed by DuPont [14], allow one enhance the overall volume flow. Axiva followed this approach in two steps. On the laboratory scale, a throughput of 6.6 kg/h at a

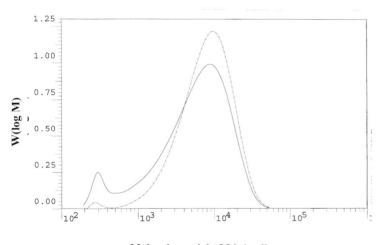

Molecular weight M [g/mol]

Fig. 6-10. Photograph of an unclogged macroscopic static mixer. No polymer precipitation occurred due to efficient premixing by means of an interdigital micromixer (top). GPC plot revealing the respective polymer fraction (bottom) [12].

pressure drop of 6.5 bar was reached in one mixer array by integration of ten mixing units. For the industrial scale, 32 such micromixers were combined in an assembly enabling an acrylate capacity of 2000 tons per year to be achieved (see Figure 6-11). In contrast, the capacity of the tube reactor could only be enlarged by classical scale-up, i.e. increasing tube diameter and, hence, reactor volume, which is associated with the classical scale-up problems.

Axiva filed a patent on this application of micromixers as a premixer and developed a flow scheme for a complete production set-up in order to offer this process variant to customers. Experience was gained with a pilot-scale plant with a capacity of 50 tons per year.

153

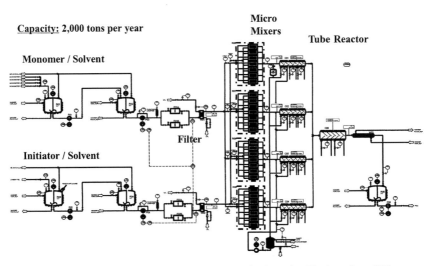

Fig. 6-11. Pre-basic design of a plant for acrylate polymerization using 32 micromixers [12].

6.4 Ketone Reduction Using a Grignard Reagent in Micromixers

Merck, Darmstadt, Germany, aimed at using microreactors for better process control and an increase in process safety [1]. In the latter case, processing in explosive regimes, handling of toxic substances and reduction of critical emissions were identified as important issues from an industrial perspective. Regarding process control, highly exothermic and endothermic reactions were supposed potentially to benefit from miniaturization. The same holds for fast reactions and temperature sensitive reactions, i.e. having a notably decreased performance already at small deviations from a set value. Finally, the analysis predicted that microreactors could be profitably used for small capacity production.

Improvement of Process Control for a Fine-Chemical Process
The latter features, referring to improvement of process control, were evaluated by a study employing a ketone reduction using a Grignard reagent as a candidate reaction. This liquid phase reaction is part of an existing Merck multi-step production process, yielding a fine chemical as final product. Regarding kinetics, this metallo-organic reaction is a fast process, being completed within seconds. However, due to high exothermicity, a relatively long operation time was required in order to guarantee sufficient heat transfer. For the large-scale batch process, this amounts to several hours, while in the laboratory only half an hour is needed.

154

By means of an interdigital micromixer (the same device as described in Section 6.3), a continuous flow process could be established, enabling the reaction time to be decreased to a few seconds. The fast mixing process allowed the reaction to be started immediately, hence to performing it under kinetically controlled conditions. Although no cooling channels were integrated in the mixing device, calculations confirmed that heat transfer through conduction via the construction material is already sufficient. Assuming operating conditions, a temperature increase of only a few kelvin was predicted, which was assumed to be tolerable.

An experimental set-up was realized by fluidic connection of three piston pumps for the two reactants and the solvent, the micromixer embedded in a cooling bath and a storage vessel. The reactants were separately premixed with a solvent and introduced in the micromixer. Within the micromixer, these solutions were contacted in an interdigital configuration consisting of microchannels of 40 μm width. The mixture was transferred from the micromixer to the storage vessel.

Statistical Approach for Yield Optimization
In the course of an experimental program, based on a statistical design of experiments, yield was monitored as a function of the process temperature and the ratio of reactants. At a flow rate of 400 ml/h, a maximum yield close to 100 % at a temperature of minus 10 °C and a ratio of Grignard reactant to ketone of about 1.8 was observed (see Figure 6-12). A steep decrease of yield at higher or lower reactant ratios resulted. A similar trend, although less pronounced, was given for the influence of temperature. Increasing the flow rate to 2.0 l/h results in an improved behavior. Only a small temperature dependence of the yield was detected, ranging from 90 to 100 % within a range of reactant ratios of about 1.6 to 2.0.

Hence high yields can be obtained at high process temperatures on increasing the volume flow. These results are in accordance with conclusions on the hydrodynamic behavior of the micromixer gained in experiments using a model reaction to characterize mixing. Thus, the increase in mixer performance may either be related to a better flow distribution of the ten mixing units or an increase in mixing quality of an individual unit [13].

The experiments concerning ketone reduction revealed that the micromixer is a precise measuring tool for process development. This clear identification of potential process improvements was so stimulating that Merck planned to establish a continuous flow production process based on such small mixing devices. For this purpose, it was decided to use laboratory-developed minimixers because during the experiments it turned out that precipitation occurred in and on the micromixing element. Whilst this did not considerably impair the laboratory investigations, it may affect the process liability of a continuous production process.

Pilot-Scale Experiments
In pilot-scale experiments, process conditions were further optimized and thereafter used for an automated production process. Five minimixers were operated in parallel, connected to a tank of several hundred liters volume. It turned out that flow manifolding to these individual units with a capacity of 3×10^{-5} m^3/s each is an important issue.

155

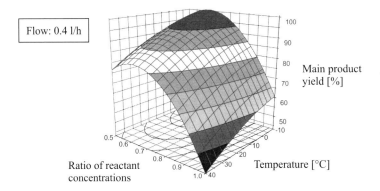

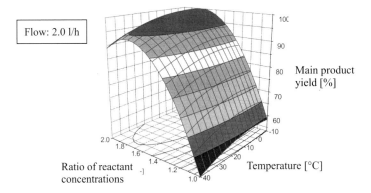

Fig. 6-12. Finding optimized process parameters for a metallo-organic reaction based on a statistical approach [1].

A yield of about 92 % was obtained, which is still higher than the laboratory-scale batch process (88 %) and considerably larger than the former batch process in production (72 %). The best performance was reached in the microreactor, amounting to a yield of 95 % (see Table 6-1). In addition, the residence time could be decreased by orders of magnitude from about 5 h to less than 10 s. This is due to replacement of the slow dropping process of one reactant into the other by fast continuous contacting. Further process benefits of the micro- and minimixers were gained because of the higher reaction temperature applicable, thereby decreasing the technical expenditure for cooling equipment and saving energy costs.

While in the micromixer, having a surface-to-volume ratio of about 10,000 m²/m³, iso-thermal conditions could be applied, this was not achievable in the 6000 l vessel of the batch process, the corresponding figure being limited to 4 m²/m³. Using a 0.5 l laboratory flask and the minimixers, with a specific surface of 80 and 4000 m²/m³, respectively, resulted in an intermediate type of behavior.

Table 6-1. Comparison of reactor performance with regard to ketone reduction for micromixer, minimixer, lab- and large-scale process [1].

Reactor type	Temperature [°C]	Residence Time	Yield [%]
0.5 l Flask	−40	0.5 h	88
Laboratory Microreactor	−10	< 10 s	95
Pilot-Scale Microreactor	−10	< 10 s	92
Production-Scale Stirred Vessel	−20	5 h	72
Production-Scale Minireactors (5)	−10	< 10 s	92

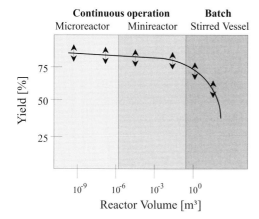

Fig. 6-13. Decrease of reactor performance as a function of characteristic internal dimensions [1].

In Figure 6-13, a general plot of the yield, as a measure of reactor performance, versus the reactor volume is given. Only a slight decrease is found on increasing the reactor volume from 10^{-9} to 10^{-3} m^3, referred to as the micro- and mini-range. A further increase in volume, also accompanied by a change from continuous to batch processing, results in a considerable performance decrease.

Concluding Remarks

On the basis of this result, the Merck researchers came to the conclusion that microreactors are suitable special-type tools for process development. They were able to carry out a statistical analysis of reactor performance in a short time and to transfer these results quickly to standard components, needing only a minimum amount of chemicals. In case of ketone reduction, a development time of only 18 months was needed from first idea until establishing the process in production, which meanwhile has been successfully running since August 1998 (see Figure 6-14).

Fig. 6-14. Photograph of a Merck production plant carrying out a metallo-organic reaction using five minimixers [1].

6.5   Laboratory-Scale Organic Chemistry in Micromixer/Tube Reactors

Researchers of TU Ilmenau (Germany) and Merck, Darmstadt (Germany), jointly developed and exploited microreaction systems based on a sequence of micromachined mixers connected by special adapters to commercial tubes of small inner diameter. Moreover, they developed an integrated silicon wafer system combining the above-mentioned component assembly within one device. First investigations were mainly aimed at proving the feasibility of reaction performance of typical organic reactions. In addition, potential benefits in terms of conversion and selectivity compared with standard laboratory equipment, e.g. stirred flasks, were identified for some reactions. As a long-term research goal, these results open the gate to a combinatorial chemistry tool for parallel organic synthesis of organic intermediates and drugs [2].

Silicon Mixer/Tube Reactor
A major driver for selecting silicon as construction material was, besides its sufficient chemical stability, the utilization of its high thermal conductivity for control of process temperature. While the micromixer served as a tool capable of superior micromixing performance, commercial tubes and a microchannel system within the integrated silicon reactor acted as a delay loop. In order to connect the microreactor with standard fluidic peripherals, e.g. piston pumps, special connectors for fluid inlet and outlet were developed, which easily allow fitting to standard equipment.

As a mixing unit, a micromixer based on a split−recombine principle, consisting of five fork-like elements [15], was utilized (see detailed description in Section 3.7.2). In order to check flow equilibration of both feeds, volume flow was monitored at constant pressure loss for introducing each liquid separately as well as jointly feeding both liquids. A small difference in volume flow for the single streams was attributed to the flow within the entrance channels, slightly varying in length from fluid connector to the first mixing element. The volume flow of both liquids was found to be higher than the single flows.

A 45 m long meandering channel served as a delay loop in the integrated silicon reactor for variation of residence times up to 60 min. The same holds for the commercial HPLC capillaries. The whole device consisted of three 4 inch silicon wafers, one being required to cover the meandering channel.

Microfabrication
Microfabrication of the full wafer microreactor was achieved by means of standard photo-lithographic processes. The process series included the deposition of masking materials, resist coating, UV irradiation, deposition of an etching mask layer, removal of resist, aniso-tropic wet chemical etching, removal of mask layer and anodic bonding.

The analysis of the correlation of volume flow to pressure loss revealed, as to be expected, an extreme decrease in the flow range achievable for the mixer structures when connected to the long delay loop system. The flow resistance increased by a factor of about 40.

Two types of detachable fluid connectors were manufactured by injection molding. An excenter actuated connector was developed for low and medium pressure applications, while a screw mounted tube connector was designed to withstand high pressure, e.g. exceeding 5 bar. Both connectors could be mounted by simple mechanical procedures.

Synthesis and Desymmetrization of Thiourea
In the framework of an experimental program, so far only the micromixer/commercial tube reactors were employed. Starting from the synthesis of a substituted thiourea from phenyl isothiocyanate and cyclohexylamine at 0 °C, the feasibility of performing a nearly sponta-neous reaction could be shown. A single mixing device connected to a stainless steel tube of about 10 m length and 0.25 mm diameter was used.

Further studies related to the desymmetrization of thioureas showed that for the diphenyl thiourea/cyclohexylamine system reasonable reaction rates and conversions could be achieved (see Figure 6-15). It is notable that the temperatures applied of up to 91°C exceed the boiling point of the solvent acetonitrile.

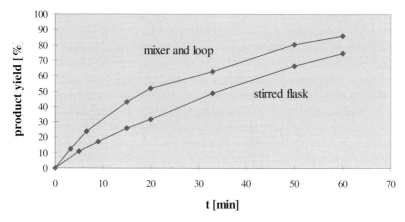

Fig. 6-15. Product yield as a function of time for desymmetrization of diphenyl thiourea using cyclohexylamine, carried out in the micromixer/tube reactor and a stirred flask [2].

Suzuki Reaction

As a second example, the Suzuki reaction, a C−C coupling reaction with palladium as catalyst, was investigated in the micromixer/tube reactor. The reaction of 3-bromo benzaldehyde and 4-fluoro phenylboronic acid could clearly show benefits in terms of conversion for the continuous flow microreactor when compared with results obtained in a laboratory flask (see Figure 6-16). This was explained by temperature and concentration profiles within the flask, representing a real stirred tank, whereas both parameters are nearly uniform in the microreactor, thereby being a model for an ideal stirred tank.

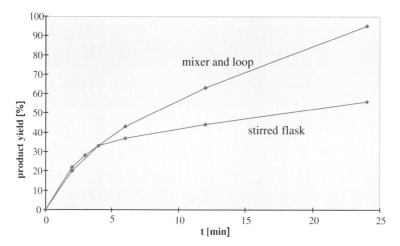

Fig. 6-16. Product yield as a function of time for the Suzuki reaction of 3-bromo-benzaldehyde and 4-fluoro-phenylboronic acid, carried out in the micromixer/tube reactor and a stirred flask [2].

Wittig−Horner−Emmons Reaction

In a further investigation in the microreactor, 4-methoxy benzaldehyde and methyl diethoxy phosphono acetate were converted by means of the Wittig−Horner−Emmons reaction to yield olefins. The microreaction system consisted of two mixers, for deprotonation of the phosphonate and introduction of the aldehyde, connected to an HPLC capillary of 0.8 m length and 0.25 mm diameter. Again, the continuous flow system turned out to be superior to the laboratory benchmark.

Hydrogenation of Methyl Cinnamate

Finally, the hydogenation of methyl cinnamate to methyl 3-phenylpropionate was investigated. Hydrogen and organic reactant were mixed in the micromixer and fed to a commercial Merck Superformance® HPLC column of 100 mm length and 5 mm inner diameter, which was used as a hydrogenator. This column contained palladium on alumina as a hydrogenation catalyst. HPLC analysis confirmed that the hydrogenated product was exclusively formed.

Concluding Remarks

The investigations of the researchers at Merck and TU Ilmenau clearly showed the high potential of small continuous flow systems regarding improvements of chemical engineering of standard organic reactions. These systems seem to provide, for many processes, an interesting alternative to existing batch technology, mainly based on stirred flasks. The combination of microstructured devices with standard laboratory equipment is, in view of the technical expenditure, a straightforward approach to a successful industrial implementation of microreaction technology.

As a unique advantage of the microsystem it turned out that inert gas purging metalloorganic mixtures, which are moisture or oxidation sensitive, is no longer necessary, due to the tight separation of reaction volumes from the environment. The results gained will turn out to be even more powerful when transferred into a combinatorial concept, either based on parallel operation of many identical devices or rapid serial variation of processes and process conditions within one device.

6.6 Dushman Reaction Using Hydrodynamic Focusing Micromixers and High-Aspect Ratio Heat Exchangers

Reseachers at the Massachusetts Institute of Technology (MIT), Cambridge (U.S.A.) developed a microreactor concept for the performance of fast, highly exothermic reactions yielding hazardous speciality chemicals [16]. As an interesting class of such chemicals organic peroxides were identified, being labile products that can undergo explosive decomposition. Organic peroxides are used for initiation of radical polymerization reactions, yielding standard polymers, e.g. polystyrene or poly(vinyl-chloride). Since the initiator is required in much lower concentrations than the monomer, typically by a factor of 1000, even small-scale production in a microreactor can potentially meet the requirements in terms of throughput of a large-scale industrial process.

Mixing Nozzle/Tube Heat Exchange Microreactor
The design of the microreactor had two characteristic features, referring to the mixing and heat exchange regions (see Figure 6-17). The micromixer comprises a number of inlet channels of 50 μm width, shaped like nozzles, arranged on a semicircle on one side of a mixing channel. These nozzles are alternately fed with two liquids from two types of parallel channels of different length which are each connected to a separate feed region consisting of inlet borings arranged on a semicircle. The multi-entrance feed through the nozzles results in a multilamination of the inlet streams. In order to enhance mixing, this multilaminated layer structure is focused by reducing the overall channel width from about 250 μm to 50 μm.

At the end of this focusing region reaction is initiated by a heat exchange channel system. In this region, the reaction channel of a constant width of 50 μm is surrounded by a double channel heat exchanger configuration, i.e. four heat exchange and one reaction channels are guided in parallel. The heat exchange channels are 50 μm wide and 350 to 500 μm high. The heat transfer coefficient for this configuration was determined to be 1445 W/m^2 K. The maximum throughput of one device was about 500 ml/h limited by the back-pressure of standard syringe pumps. The total size of the microreactors amounts to 24 mm x 29 mm.

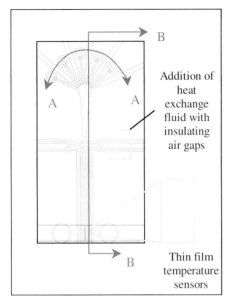

Fig. 6-17. Microreaction system consisting of a nozzle-type mixing component, using multilamination and hydrodynamic focusing, and a two-channel, countercurrent flow heat exchanger [16].

Microfabrication

As in the case of the gas phase microreactor of MIT [17], a silicon-on-insulator (SOI) wafer was microstructured, consisting of a silicon platelet of varying thickness and a 1 μm silicon oxide layer. Apart from standard lithographic techniques, deep reactive ion etching (DRIE) was employed. The wafer structure was sealed by anodic bonding with Pyrex glass. The device could be additionally equipped with temperature sensors by means of thin film techniques.

Mixing Efficiency and Dushman Reaction

For determination of mixing efficiency, an acid–base reaction visualized by a color change due to a pH indicator was employed. The existence of a hydrodynamically focused multilaminated layer structure could be confirmed by microscopy analysis. Measuring UV absorption allowed mixing efficiency to be monitored as a function of the flow rate. Thereby, a mixing time of about 10 ms was derived. The experimentally gathered data are in accordance with theoretical descriptions of species transport as evidenced by CFD simulation.

The Dushman reaction, the formation of iodine by an acid-induced redox process, was used to evaluate reactor performance. As to be expected, the product yield increased on reducing flow rates (see Figure 6-18), i.e. increasing residence time. Whilst this result underlines again the importance of micromixing for completion of reaction, so far no information about the impact of the micro heat exchange structure has been given, which certainly remains to be addressed in future investigations.

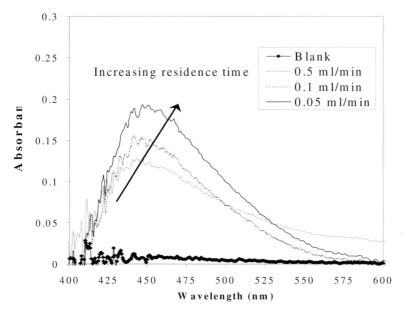

Fig. 6-18. Feasibility of carrying out the Dushman reaction in a microreaction system [16].

$$5 \, I^{\ominus} + IO_3^{\ominus} + 6H^{\oplus} \longrightarrow 3 \, I_2 + 3 \, H_2O$$

Concluding Remarks

The concept of the MIT group is similar to that of a gas phase microreactor design, already published [17] by the same authors, with respect to system complexity. It includes a micromixer and heat exchanger structure as well as miniaturized on-line sensors. Similarly to a microreactor concept for liquid/liquid reactions [5] deep microchannels with large contact surfaces are employed to guarantee efficient heat exchange. The present study already confirmed the feasibility of the concept. The real potential of this liquid microreactor will certainly become evident in subsequent research studies.

6.7   Synthesis of Microcrystallites in a Microtechnology-Based Continuous Segmented-Flow Tubular Reactor

Although precipitation phenomena are generally prohibitive during operation in microreactors, a controlled synthesis of particles much smaller than the microchannel size is far from being a utopia. This holds in particular if residence times of such particles can be kept sufficiently short and if microreactor wall roughness and surface properties can be adjusted.

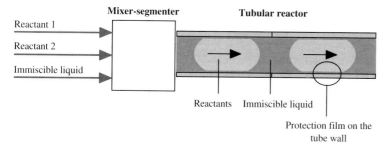

Fig. 6-19. Flow scheme of a continuous segmented-flow tubular reactor (SFTR) (by courtesy of Donnet, EPFL Lausanne).

First scouting experiments on the precipitation of microcrystallites in micromixers were performed by Schmidt and Donnet at the EPFL Lausanne (Switzerland) [18, 19]. The micromixers, fabricated by the Institut für Mikrotechnik Mainz (Germany), were part of a more complex system, referred to as a continuous segmented flow tubular reactor (SFTR) [20]. This system comprises two mixing functions for generation of seed crystals by addition of precipitating agents to a solution and separation of these reaction volumes by segmenting them with kerosene plugs (see Figure 6-19).

The first process, the seed crystal generation, is sensitive to micromixing phenomena, and thus potentially should benefit from operating using a miniaturized mixer providing a dead-zone free processing. Hence, the first mixer of the SFTR system was replaced by an interdigital micromixer (see Section 3.7.1), the other components being retained.

Copper Oxalate Precipitation
First experiments were carried out using the precipitation of copper oxalate as a model system. Prior experiments using conventional components proved the formation of a small number of larger particles in addition to the desired microcrystalline fraction with an average diameter of about $2.9 \pm 1.0 \, \mu m$. Using the interdigital micromixer, a more homogeneous particle size distribution could be obtained (see Figure 6-20).

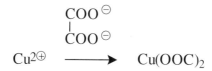

Concluding Remarks
These first feasibility experiments are stimulating further research, but pose questions as well. It has to be proven that a reliable operation can be achieved at high throughputs. Certainly, micromixers made of optimized materials and adjusted surface roughness have to be provided. Moreover, a detailed knowledge of flow patterns within the micromixer has

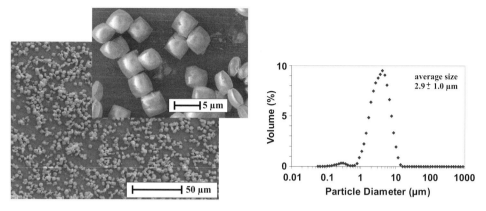

Fig. 6-20. Scanning electron micrograph of microcrystallites of copper oxalate obtained by processing in an interdigital micromixer as part of a SFTR system (left). Particle size distribution of the corresponding microcrystallites [19].

to be gained by means of fluid dynamic simulations. Nevertheless, the motivating results concerning copper oxalate precipitation add suspension formation as a new research field for micromixer investigations to the established activities concerning gas/liquid dispersions and emulsions.

6.8   Electrochemical Microreactors

6.8.1   Synthesis of 4-Methoxybenzaldehyde in a Plate-to-Plate Electrode Configuration

Benefits and Limitations of State-of-the-Art Electrochemistry
Electrochemical processes employing current as reactive agent have a number of advantages compared with conventional liquid phase processes based on utilization of various chemical compounds. In particular, oxidation and reduction processes benefit from continuous variation of reactivity by means of regulation of electric potential. This has clear benefits in terms of product separation and process flexibility. In addition, reactions can be carried out at room temperature.

However, several limitations of current electrochemical process technology have so far hindered its widespread use in synthetic chemistry. To mention only one drawback, current cells still suffer from inhomogeneities of the electric field. Microreaction systems which can be geometrically precisely fabricated may help to overcome some of these limitations. Apart from field homogeneity, such electrochemical microreactors generally should exhibit a uniform flow distribution, i.e. without any dead zones, and efficient heat transfer.

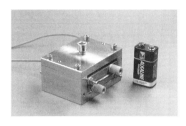
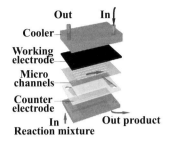

Fig. 6-21. Photograph of a mounted electrochemical microreactor utilizing thin layer technology via a plate-to-plate configuration and schematic of its assembly [21].

Plate-to-Plate Electrode Microreactor

Researchers at the Institut für Mikrotechnik Mainz (Germany) developed an electrochemical microreactor concept [21, 22] based on the utilization of microstructuring techniques for thin layer technology [23], so far being applied in analytics. The main component of the stacked-platelet-type microreactor is a microchannel layer embedded between a working and a counter-electrode, referred to as a plate-to-plate configuration (see Figure 6-21). Furthermore, an integrated cooling element served for efficient control of reaction temperature.

The microstructured platelets were realized by etching of polymer foils and laser cutting techniques. Microchannels of 800 µm width, 25 µm depth and 64 mm length were operated at volume flows of 6 ml/h for a pressure drop of 1.1 bar. Due to the small diffusion layer thickness, fast mass transfer between the electrodes is achievable. The ratio of electrode surface normalized by cell volume amounts to 40,000 m^2/m^3. This value clearly exceeds the specific surface areas of conventional mono- and bipolar cells of 10 to 100 m^2/m^3.

Electrosynthesis of 4-Methoxybenzaldehyde

The electrosynthesis of 4-methoxybenzaldehyde from 4-methoxytoluene by means of direct anodic oxidation is performed on an industrial scale [24]. Via an intermediate methyl ether derivative the corresponding diacetal is obtained which can be hydrolyzed to the target product [25]. The different types of products–ether, diacetal, aldehyde–correspond to three distinct single oxidation steps.

167

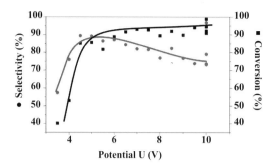

Fig. 6-22. Feasibility of carring out the direct anodic oxidation of 4-methoxytoluene to 4-methoxybenzaldehyde in a plate-to-plate microreactor, as evidenced by high conversion and selectivities for proper setting of electric potential [21].

Carrying out the oxidation of 4-methoxytoluene, a constant current density of 79 mA/cm^2 was employed using 0.1 M KF as conducting salt (see Figure 6-22). Both, the conversion and selectivity of this reaction increased rapidly at low values of electric potential. A potential of about 4.5 V provides the best results, namely a selectivity and conversion of about 86 % and 88 %, respectively. At higher potentials, conversion slightly increases, whereas selectivity decreases due to side reactions. By means of HPLC analysis, it could be shown that all three types of products were formed during reaction in the microreactor. For a typical experiment, the main fraction was composed of the diacetal, the second largest fraction consisted of the aldehyde, and the monoether was generated to a smaller extent.

A new type of processing could be realized by omitting the conductive salt, thereby simplifying product separation and reducing operational costs. The feasibility of this approach could be proven, e.g. by determining selectivities of up to 90 %, although the conversion dropped notably compared with processing with conductive salt (in this case, only 0.01 M KF was employed). In particular, the current efficiency (up to 98 %) within the electrolyte is much larger when using no conductive salt (see Figure 6-23). This is a clear indication of a reduction of side reactions for this type of processing. In addition, a temperature increase allows the conversion of the salt-free process to be increased while maintaining constant selectivity.

Concluding Remarks
At present, descriptions of applications using electrochemical microreactors are still rare. However, there is no denying the fact that notable benefits due to miniaturization have to be expected theoretically. Since some of the current limitations of electrochemistry actually may be overcome by using microstructured devices, this could put a new complexion on the general and widespread utilization of this technology. These predictions need to be proven. Carrying out more experiments which may be accomplished by potential finding of unexpected phenomena, e.g. axial heat conduction with regard to heat transfer in miniaturized devices (see Section 4.4), will shed more light on the real performance achievable.

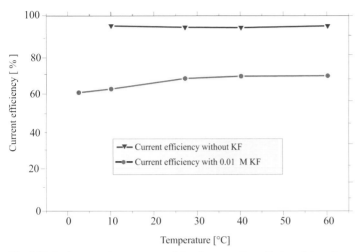

Fig. 6-23. Performance improvement of the direct anodic oxidation of 4-methoxy toluene in terms of current efficiency by operating without a conductive salt in a plate-to-plate microreactor [21, 22].

6.8.2   Scouting Potentiodynamic Operation of Closed Microcells

Thin Film Cells

Long time before electrochemical microreactors were realized, small-sized electrochemical cells have been described. In the seventies, flow-through systems with small channels separated by a working and counter electrode, termed thin film cells [26, p.406], were developed. The distance between the working and counter electrodes, i.e. the channel height, typically amounted to 10 to 100 µm. These systems benefit from a simpler theoretical description of basic transport phenomena as well as from optimized experimental performance, e.g. resulting in enhanced space-time yields and reduced energy consumption.

Open Microcells

Single open microcells, which e.g. allow an observation via microscope, were developed just recently. One possible type is given by the placement of a drop of electrolyte solution on a plane working electrode, partially covered by a photoresist [27]. Small reference and counter electrodes are inserted into the drop by means of optical control and an accurate micropositioning system. Another variant refers to the fixation of an electrolyte drop on a plane working electrode by means of a gold capillary, which is internally filled with electrolyte. The capillary is used as counter electrode and, in addition, the electrolyte inside acts as a salt bridge to connect the reference electrode [27].

Closed Microcells

In order to combine the advantages of thin film cells and open microcells, the group of Schultze at the Universität Düsseldorf (Germany) realized a closed microcell [27]. This device is a flow-through system of defined, small internal dimensions which also allows observation and micropositioning of small-sized electrodes. These microelectrodes are manufactured by specialized, often individually developed methods. They comprise a glass capillary into which a wire, e.g. made of platinum or silver, is inserted. A tight system is achieved by means of welding or glueing followed by polishing of the front part of the capillary. For one typical electrode, the outer diameter of the glass capillary amounts to 20 µm, while the inner diameter of the platinum wire is 10 µm. Accordingly, the corresponding surface area of the platinum wire can be set as low as 10^{-6} mm^2. The such modified capillary is fixed in a cap which contains an electrical connector.

A microcell of 10 mm x 10 mm x 5 mm size was assembled from conventional PMMA cube set on five 100 µm thin microstructured Foturan$^{®}$ layers. These glass layers were realized by a combination of photoetching and thermal treatment as described in Section 2.7. The transparent glass material allowed an observation of the electrolyte chamber of the flow-through system in the center of the microcell. This chamber contains an opening at the bottom which is partially encaved by a seal of silicon rubber. This seal can be placed on plane working electrodes of various materials, thereby guaranteeing a fluid tight connection. Counter and reference electrodes are introduced into the electrolyte chamber by insertion of small micro-electrode wires through connectiong plugs. These wires are modified further as desribed above. In addition, in- and outlet tubes of 300 µm internal diameter are placed on top of the microcell. The experimental set-up contained, apart from the microcell, a potentiostat connected to the electrodes, a X-Y-Z positioning system, a microscope, and a micropump connected to an electrolyte reservoir.

Potentiodynamic Operation
The size of the electrolyte chamber is defined by the geometry of the rubber seal. In a first scouting experiment, an oval hole of 700 µm long axis and about 350 µm small axis was utilized. Thereby, the surface of a titanium electrode was locally oxidized by a potentiodynamic scan, serving as an example for surface modification. A second experiment demonstrated the feasibility of performing a microelectrochemical analysis by a potentiodynamic scan, a current-voltage plot, on a platinum electrode using aerated sulphuric acid as electrolyte.

Concluding Remarks
The realization of a closed microcell is an example of a successful implementation of microtechnique-made devices into existing small-sized chemistry. Thereby, start-up time to carry out experiments is reduced and, moreover, start-of-the-art technical know-how can be utilized. The application of glass as construction material is ultimately needed, since microscopy observation is crucial in case of miniaturized electrochemical cells. Future measurements have to show the advantages of using closed microcells for analytical purposes or chemical syntheses, e.g., compared to thin film cells and open microcells.

6.9 References

[1] Krummradt, H., Kopp, U., Stoldt, J.; *"Experiences with the use of microreactors in organic synthesis"*, in Ehrfeld, W. (Ed.) *Microreaction Technology: 3rd International Conference on Microreaction Technology, Proceedings of IMRET 3*, pp. 181–186, Springer-Verlag, Berlin, (2000).

[2] Schwesinger, N., Marufke, O., Qiao, F., Devant, R., Wurziger, H.; *"A full wafer silicon microreactor for combinatorial chemistry"*, in Ehrfeld, W., Rinard, I. H., Wegeng, R. S. (Eds.) *Process Miniaturization: 2nd International Conference on Microreaction Technology; Topical Conference Preprints*, p. 124, AIChE, New Orleans, USA, (1998).

[3] Antes, J., Krause, H. H., Löbbecke, S.; *"Kombinatorischer Einsatz von Mikrostrukturen zur Auslegung und Optimierung von Polymersynthesen"*, in Proceedings of the "DECHEMA Symposium: Kombinatorische Katalyse- und Materialforschung", 21 June, 1999; p. P 15; Frankfurt, Germany.

[4] de Bellefon, C., Tanchoux, N., Caravieilhes, S., Hessel, V.; *"New reactors for liquid-liquid catalysis"*, in Proceedings of the "Entretiens Jaques Cartier", 6–8 Dec., 1999; Lyon.

[5] Wörz, O., Jäckel, K. P., Richter, T., Wolf, A.; *"Microreactors, a new efficient tool for optimum reactor design"*, in Ehrfeld, W., Rinard, I. H., Wegeng, R. S. (Eds.) *Process Miniaturization: 2nd International Conference on Microreaction Technology, IMRET 2; Topical Conference Preprints*, pp. 183–185, AIChE, New Orleans, USA, (1998).

[6] Wörz, O., Jäckel, K. P.; *"Winzlinge mit großer Zukunft – Mikroreaktoren für die Chemie"*, Chem. Techn. **26**, 131 (1997) 130–134.

[7] Ehrfeld, W., Hessel, V., Haverkamp, V.; *"Microreactors"*, Ullmann's Encyclopedia of Industrial Chemistry, Wiley-VCH, Weinheim, (1999).

[8] Wolf, A., Ehrfeld, W., Lehr, H., Michel, F., Richter, T., Gruber, H., Wörz, O.; *"Mikroreaktorfertigung mittels Funkenerosion"*, F & M, Feinwerktechnik, Mikrotechnik, Meßtechnik, pp. 436–439, 6 (1997).

[9] Richter, T., Ehrfeld, W., Wolf, A., Gruber, H. P., Wörz, O.; *"Fabrication of microreactor components by electro discharge machining"*, in Ehrfeld, W. (Ed.) *Microreaction Technology, Proc. of the 1st International Conference on Microreaction Technology*, pp. 158–168, Springer-Verlag, Berlin, (1997).

[10] Richter, T., Ehrfeld, W., Hessel, V., Löwe, H., Storz, M., Wolf, A.; *"A flexible multi component microreaction system for liquid phase reactions"*, in Ehrfeld, W. (Ed.) *Microreaction Technology: 3rd International Conference on Microreaction Technology, Proceedings of IMRET 3*, pp. 636–644, Springer-Verlag, Berlin, (2000).

[11] Richter, T., Ehrfeld, W., Gebauer, K., Golbig, K., Hessel, V., Löwe, H., Wolf, A.; *"Metallic microreactors: components and integrated systems"*, in Ehrfeld, W., Rinard, I. H., Wegeng, R. S. (Eds.) *Process Miniaturization: 2nd International Conference on Microreaction Technology, IMRET 2; Topical Conference Preprints*, pp. 146–151, AIChE, New Orleans, USA, (1998).

[12] Bayer, T., Pysall, D., Wachsen, O.; *"Micro mixing effects in continuous radical polymerization"*, in Ehrfeld, W. (Ed.) *Microreaction Technology: 3rd International Conference on Microreaction Technology, Proceedings of IMRET 3*, pp. 165–170, Springer-Verlag, Berlin, (2000).

[13] Ehrfeld, W., Golbig, K., Hessel, V., Löwe, H., Richter, T.; *"Characterization of mixing in micromixers by a test reaction: single mixing units and mixer arrays"*, Ind. Eng. Chem. Res. **38**, 3 (1999) 1075–1082.

[14] Lerou, J. J., Harold, M. P., Ryley, J., Ashmead, J., O'Brien, T. C., Johnson, M., Perrotto, J., Blaisdell, C. T., Rensi, T. A., Nyquist, J.; *"Microfabricated mini-chemical systems: technical feasibility"*, in Ehrfeld, W. (Ed.) *Microsystem Technology for Chemical and Biological Microreactors*, Vol. 132, pp. 51–69, Verlag Chemie,, Weinheim, (1996).

[15] Schwesinger, N., Frank, T., Wurmus, H.; *"A modular microfluid system with an integrated micromixer"*, J. Micromech. Microeng. **6**, pp. 99–102 (1996).

[16] Floyd, T. M., Losey, M. W., Firebaugh, S. L., Jensen, K. F., Schmidt, M. A.; *"Novel liquid phase microreactors for safe production of hazardous specialty chemicals"*, in Ehrfeld, W. (Ed.) *Microreaction Technology: 3rd International Conference on Microreaction Technology, Proceedings of IMRET 3*, pp. 171–180, Springer-Verlag, Berlin, (2000).

[17] Jensen, K. F., Hsing, I.-M., Srinivasan, R., Schmidt, M. A., Harold, M. P., Lerou, J. J., Ryley, J. F.; *"Reaction engineering for microreactor systems"*, in Ehrfeld, W. (Ed.) *Microreaction Technology, Proceedings of the 1st International Conference on Microreaction Technology; IMRET 1*, pp. 2–9, Springer-Verlag, Berlin, (1997).

[18] Donnet, M.; *"Results of EPFL, Lausanne; unpublished"*, (1999).

[19] Löwe, H., Ehrfeld, W., Hessel, V., Richter, T., Schiewe, J.; *"Micromixing technology"*, in Proceedings of the "4th International Conference on Microreaction Technology, IMRET 4", pp. 31–47; 5–9 March, 2000; Atlanta, USA.

[20] Jongen, N., Lemaître, J., Bowen, P., Hofmann, H.; *"Oxalate precipitation using a new tubular plug-flow reactor"*, in Proceedings of the "5th World Congress of Chemical Engineering", 14–18 July, 1996; pp. 31–36; San Diego.

[21] Löwe, H., Ehrfeld, W., Küpper, M., Ziogas, A.; *"Electrochemical microreactors: a new approach in microreaction technology"*, in Ehrfeld, W. (Ed.) *Microreaction Technology: 3rd International Conference on Microreaction Technology, Proceedings of IMRET 3,* pp. 136–156, Springer-Verlag, Berlin, (2000).

[22] Löwe, H., Ehrfeld, W.; *"State of the art in microreaction technology: concepts, manufacturing and applications"*, Electrochim. Acta **44,** pp. 3679–3689 (1999).

[23] Woodard, F. E., Reilley, C. N.; *"Thin layer cell techniques"*, in Yeager, V., Bockris, J. O. M., Conway, P. P., Sarangabani, S. T. (Eds.) *Comprehensive Treatise of Electrochemistry,* Vol. 9, pp. 353–392, Plenum Press, New York, (1984).

[24] Degner, D., Barl, N., Siegel, H.; *"Elektrochemische Herstellung von in 4-Stellung substituierten Benzaldehyddialkylacetalen"*, DE 28 48 397, (08.11.1978); BASF.

[25] Wendt, H., Bitterlich, S.; *"Anodic synthesis of benzaldehydes-I. Voltammetry of the anodic oxidation of toluenes in non-aqueous solutions"*, Electrochim. Acta **37,** p. 1951, 11 (1992).

[26] Bard, A., Faulkner, W.; *Electrochemical Methods: Thin Layer Electrochemistry,* Wiley, New York.

[27] Vogel, A., Schultze, J. W., Küpper, M.; in Ehrfeld, W., Rinard, I. H., Wegeng, R. S. (Eds.) *Process Miniaturization: 2nd International Conference on Microreaction Technology; Topical Conference Preprints,* pp. 323–328, AIChE, New Orleans, USA, (1998).

7 Microsystems for Gas Phase Reactions

A number of microreactor applications refer to gas phase syntheses since a number of reactions of this class virtually ideally meet the requirements needed to benefit from miniaturization, as defined in Chapter 1: many gas phase processes, especially fast high-temperature reactions, are controlled by mass diffusion rather than by reaction kinetics. Since the overwhelming majority of these reactions are also heterogeneously catalyzed, i.e. conversion occurs at a gas/solid interface, the respective mass transport limitations are even boosted by facing a multiphase system.

The same scenario holds for heat transfer, thereby rendering benefits from processing in a microsystem for reactions with, e.g., huge heat releases due to a large reaction enthalpy. Finally, gas phase reactions are usually characterized by a relatively high process complexity, being composed of a net of elemental reactions leading to a product as well as by-products by means of side and follow-up processes. The extent of product formation with respect to by-products, in other words the process selectivity of a gas phase process, is significantly impacted by transport limitations mentioned above.

7.1 Catalyst Supply for Microreactors

Nearly all gas phase processes inevitably require active catalysts. Commercially, pellets coated with small amounts of spread catalytic materials are applied as tube fillings in fixed bed reactors. Moreover, fluidized-bed reactors utilize catalytically active powders which are finely dispersed within the reactor volume. Finally, gauzes of catalytic material, most often noble metals, are commonly applied for reactions of short residence time. For a number of reasons, these concepts cannot be applied when using microreactors. Most prominent, the irregular packing of powders or small pellets would abolish the advantages of a uniform temperature and concentration profile, e.g. may favor the formation of hot spots. A similar drawback is the increase in pressure drop due to the irregular flow profile. In addition, practical aspects, concerning a defined filling of microstructures with microscale particles, hinder widespread use.

Bulk Material and Thin Film Coated Catalysts
The absence of conventional solutions led to the development of specific, if not unique, techniques of catalyst supply in microreactors. For instance, catalysts are applied as microstructured catalytic bulk material [1], by means of wet impregnation of nanoporous carrier platelets [2, 3], and by physical vapor deposition (PVD) [4] as well as chemical vapor deposition (CVD) of catalytic thin films on microstructured surfaces [5].

Anodic Oxidation, Sol–Gel Techniques, and Immobilized Nanoparticles
Recent approaches tend to focus on the deposition of nanoporous ceramic layers of defined thickness on microstructured platelets, e.g. by means of anodic oxidation (see Section 7.3.2)

5 µm ⊢⊣   300 nm ⊢⊣

Fig. 7-1. Scanning electron micrograph of an alumina layer deposited by sol-gel technique on a stainless steel platelet [7].

[6]. Deposition of catalytically active material is performed subsequently by means of impregnation or precipitation. The composition of such modified ceramic layers is similar to that of the pellets, although different in geometry. Thus, this approach combines the use of material combinations of known activity with the provision of sufficient porosity for surface enlargement of the microstructures.

Sol–gel dipping provides another technique capable of generation of nanoporous ceramic layers on microstructured surfaces. Figure 7-1 shows an alumina layer deposited by sol-gel technique on a stainless steel platelet [7]. It is evident that the surface is significantly enlarged after coating. According to the process variant used, the formation of an α-phase was expected. First analysis of the chemical composition and crystal structure by Fourier Transform Infrared (FTIR) spectroscopy- and X-ray diffraction (XRD) was performed and, in addition, posed a number of questions [8]. Nevertheless, this shows that much more research work is required to elucidate such complex material combinations.

In another study, porous coatings made of aluminum, silicon, and titanium oxides were coated by sol–gel-processes on an Fe-based alloy [9]. For an alumina layer, the specific surface area first increased upon temperature treatment after sol–gel coating and thereafter decreased, but still exceeding the uncoated stainless steel surface. At a temperature of 500 °C, the highest specific surface area of 152 m^2/m^3 was observed. A further parameter for optimization of preparation was provided by a suitable choice of precursor material. Using a *sec.*-butylate aluminum precursor, the specific surface area could be further increased up to 430 m^2/m^3. The corresponding values for the silicon and titanium oxide layers were lower. An analysis of the pore structure of these layers by means of nitrogen adsorption isotherms revealed micropores of 5 to 6 Å and mesopores of 40 Å average size.

As an alternative approach to the sol–gel technique, coatings based on immobilized nanoparticles have been mentioned [9]. One advantage of these coatings is that they bear the catalytic active material already after deposition and, consequently, need no solution post-processing, e.g. by impregnation or precipitation. These coatings were applied in a microchannel device performing methanol–steam reforming [10]. The catalyst coatings showed high activity at low residence times (see Figure 7-2).

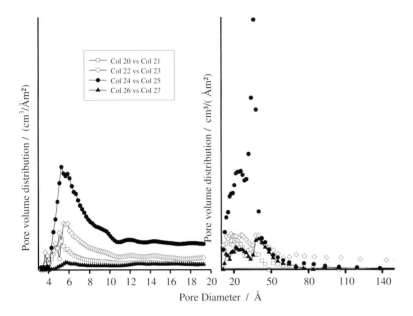

Fig. 7-2. Scanning pore size distribution according to the HK (for micropores) and BJH model (for mesopores) for various sol/gel coatings on a metallic support [9].

Slurry and Aerosol Techniques

As a third alternative to sol–gel and nano-scale coatings, slurry techniques have to be mentioned (see e.g. [11, 12]), but so far have not been applied. A distinct advantage of this technique would be the realization of much thicker catalyst carrier layers. Sol–gel coating being limited to about 1 μm, the former techniques permits the deposition of ceramic coatings up to a few hundred micrometers thickness. CFD simulations showed that mass and heat transport in porous media of this size embedded in 500 μm wide channels are still sufficient to guarantee efficient operation of the microreactor [8]. Hence, regarding productivity aspects, microreactors with a ratio of channel width to catalyst layer thickness of about 2.5 to 10 turned out to be the optimum compromise between the decrease of linear dimensions and the increase in specific catalytic volume in this case.

Aerosol techniques using metal salt solutions for catalyst deposition were applied to coat uniformly a gas phase microreactor [13]. Previous attempts to use wet chemical deposition techniques suffered from non-uniform coating of the microchannel, due to surfaces forces being mainly restricted to the channel edges. Aerosol deposited catalysts made of platinum, silver, and rhodium displayed high surface areas. The same microreactor was also coated by electroplating using a PVD seeding layer and a platinum salt solution [4].

Another important issue for future investigations is the reduction of undesired coatings on microchannels or catalyst surfaces, in particular, referring to carbon deposition, e.g. as soot layers. It was shown by a number of investigations that soot formation can be inhibited in microreactors, e.g. by carrying out fast high-temperature reactions under non-equilibrium conditions [14].

Nevertheless, it cannot be expected that such deposits will be absent for all reactions to be investigated in microreactors. It is also known that depositions, often of polymeric nature, occur for some liquid phase reactions in microreactors. In this context, new approaches aiming at controlled "roughening" of surfaces and change in surface energy in order to minimize adhesion of solid material are of notable interest [15].

7.2 Types of Gas Phase Microreactors

A classification of gas phase microreactors based on the type of supply of the catalyst material is not recommended at present. Currently, the supply technique chosen usually depends more on fabrication issues, rather than on reaction engineering aspects. Moreover, promising techniques such as the deposition of nanoscale ceramic layers as catalyst carriers are still in their infancy.

Since the majority of gas phase microreactors are complexly assembled systems, a division cannot be based on single design features, such as flow configurations comprising a certain mixing or heat transfer principle. Due to this complex assembly, it was decided to arrange gas phase microreactors in groups belonging to a certain combination of components. Therefore, the classification chosen refers to a ranking with respect to system complexity.

Relevant components for today's gas phase applications are micromixers, micro heat exchangers, catalyst structures, and on-line sensors. This definition does not include external heating sources, e.g. an oven surrounding a microreactor or heating cartridges inserted in it. In addition, macroscopic sensing equipment, e.g. thermocouples, were excluded. These parts are not integrated in the microsystem, and thus present hybrid solutions as a compromise between miniaturized and conventional approaches.

According to this definition, the following classes of current gas phase microreactors can be identified:

- microdevices with catalyst structures only
- microsystems with catalyst structures and heat exchanger
- microsystems with catalyst structures and micromixer
- microsystems with catalyst structures, heat exchanger and sensors
- complex microsystems with catalyst structures, micromixers, heat exchanger and sensors

Microdevices with catalyst structures only will be termed microchannel catalyst structures in the following.

It has to be pointed out that system complexity does not necessarily correspond to superior reactor performance. On the contrary, a number of experimental investigations have shown that microchannel catalyst structures yielded promising results. Facing issues of standardization, future work will have to elucidate how much complexity of a microeactor is ultimately needed to achieve process benefits. In other words, this will define the extent of manufacturing expenditure that can be economically tolerated.

7.3 Microchannel Catalyst Structures

Microchannel catalyst structures are similar to monolithic reactors which are used, in addition to exhaust gas burning and automotive applications, in heterogeneous catalysis [16]. This similarity especially holds for ceramic monoliths composed of straight channels, e.g. fabricated by precision engineering techniques. However, the characteristic reactor dimension of microchannels, i.e. the hydraulic diameter, is an order of magnitude smaller than that of monoliths.

A frequently used approach to realize microchannel catalyst structures is to build up a stack of microstructured platelets. The first approach described was based on a design of a cross-flow micro heat exchanger already presented in Section 4.1.1 [17]. In this concept, platelets made of a solid catalytically active material, e.g., copper, silver or rhodium, were used. Either a cross-flow design concept comprising internal cooling channels or stacks without integrated heat exchange were applied. In the latter case all channels were operated co-currently and only fed by the reactant gas, whilst heat transfer was performed from an external source, e.g. an oven. Only these externally heated components are described in the following, whereas the cross-flow devices, better being characterized as combined reactor/heat exchanger systems, will be discussed in the next chapter. In addition, recent developments based on counter flow devices will be introduced.

A relatively large variety of platelet materials are available, ranging from catalytic bulk materials, such as copper or silver, to non-catalytic carriers for catalysts, such as aluminum or various types of stainless steel. In the case of aluminum platelets, an anodic oxidation process was used to create an oxide layer with a regular pore structure in which palladium was embedded as catalytically active material [2, 6, 18, 19]. Copper platelets with oxidized surfaces [20, 21] and silver platelets [22, 23] served as bulk catalytic materials. Stainless steel platelets were subsequently coated with Al_2O_3 by means of a CVD process and platinum as catalytically active material by means of a standard wet impregnation process [24].

7.3.1 Flow Distribution in Microchannel Catalyst Reactors

Fluid dynamic aspects of stacked platelet devices like microchannel catalyst reactors were investigated in a theoretical study conducted by Wießmeier and Hönicke [19] as well as at the Universität Erlangen (Germany) and Acordis Research, Obernburg (Germany) [25]. In particular, the influence of the inlet geometries of fluid connectors of such devices on flow equipartition was determined. Based on typical dimensions of gas phase microreactors as manufactured at the Forschungszentrum Karlsruhe, it was found that vortex formation occurs at higher volume flows within the inlet region [19],[25]. These turbulent zones result in flow maldistribution between channel arrays on the microstructured platelets, e.g. inducing much higher flows in the inner platelets as compared with the outer platelets [25].

The impact of flow maldistribution on the yield of an intermediate product within a consecutive reaction was calculated. Assuming typical kinetic data, a difference in flow velocities in single channel arrays of about 20 % results in a variation of product yields by only about 3 %. Taking into account that the total yield of the microreactor is still reduced to a smaller extent due to averaging from these single yields, it can be stated that flow maldistribution, for the application outlined above, plays only a minor role for microreactor performance.

Only in the case of much higher degrees of flow maldistribution the performance of the microreactor will be notably reduced. However, it was pointed out that it is a typical feature of consecutive reactions that around the optimum flow rate their sensitivity to flow variations is not too pronounced. Other types of reactions may be much more sensitive to differences in single channel feed.

7.3.2 Partial Oxidation of Propene to Acrolein

Wießmeier and Hönicke at the TU Chemnitz (Germany) aimed at a model reaction conferring benefits of reactor performance by improved control over temperature and residence time [2]. In a series of experimental investigations, they applied a microchannel catalyst structure consisting of stacked platelets with microchannels, manufactured at the Forschungszentrum Karlsruhe (Germany). The results of this work were reported in a number of publications and in theses [19, 21].

One of the first reactions investigated in this microchannel catalyst structure was the heterogeneously catalyzed partial oxidation of propene to generate acrolein. The aim was to prevent total oxidation to yield carbon dioxide and to quench the reaction after synthesis of the thermodynamically unstable intermediate acrolein, considering the chosen process parameters.

$$\text{propene} \xrightarrow[\text{Cu}_2\text{O}]{\text{O}_2} \text{acrolein-CHO}$$

Interplay of Catalyst Composition and Selectivity

For this purpose, platelets consisting of copper with a catalytically active surface of Cu_2O were employed [20, 21]. Typical reaction conditions were characterized by a temperature range of 350 to 375 °C with nitrogen as a carrier gas. Typical concentrations of propene and oxygen were below 1 vol.-%.

One important result of these experiments, performed during the infancy of microreaction technology, was to demonstrate that a cold product gas can be heated nearly instantane-

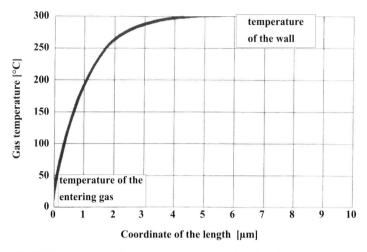

Fig. 7-3. Temperature profile along the flow axis of a microchannel catalyst structure calculated for typical process conditions of the partial oxidation of propene to acrolein [20].

ously to the reaction temperature. Figure 7-3 shows that thermal equilibrium is established after a short entrance length, amounting to 4 μm in order to reach 300 °C using propene. Therefore, channel lengths and, consequently, pressure losses can–at least in selected cases– be kept as short as possible, only being determined by the reaction kinetics.

However, in order to study reactor performance, the suitability of the propene oxidation as model reaction turned out to be limited. This limitation mainly refers to the nature of the catalyst required, namely an oxidized copper(I) surface. In this case, the surface is sensitive to changes in composition of the catalytically active centers by contact with oxygen during reaction. Hence it was observed that the propene conversion rate and selectivity changed considerably during long-term operation of the reactor. To overcome this situation, bromomethane (2.25 ppm) was used as a promoter to stabilize the composition of the metal oxide layer. An addition of this ingredient to the gas phase resulted in an increase of selectivity and only a slight decrease of conversion.

Concluding Remarks
Being one of the first investigations in the field of microreaction technology, the above experiments have to be regarded as important scouting work to utilize microchannel reactors. In particular, a potential field of application was clearly figured out, namely the preferred isolation of thermodynamically unstable intermediates, as demonstrated by the example of the oxidation of propene to acrolein.

These experiments paved the way for further experiments with model reactions using less sensitive catalyst surfaces and also stimulated further design and construction improvements of the microchannel structure which are reported below.

7.3.3 Selective Partial Hydrogenation of a Cyclic Triene

Apart from reflecting catalyst activity, the first scouting experiments (see Section 7.3.1) performed at the TU Chemnitz (Germany) already pointed out another important research issue for catalytic applications. The results clearly indicated that, at least for some applications, plane surfaces do not provide enough surface area for catalytic reactions. This stimulated the development of catalyst layers of sufficient porosity in order to enhance conversion and to realize acceptable space–time yields.

Compared with macroscopic configurations, the surface area-to-volume ratio of microchannel devices is enhanced by orders of magnitude, typically amounting to a factor of about 100 to 1000. Nevertheless, these specific interfacial areas are far from being close to values for nanoporous materials typical for many catalysts and catalyst carriers. Accordingly, equivalent surfaces of microchannel structures have to be further increased by a factor of 100 to 10,000. Such a surface enhancement cannot be achieved by means of microfabrication or standard mechanical surface roughening techniques. The same holds when employing PVD or CVD techniques.

Surface Enlargement of Catalyst Carrier by Means of Anodic Oxidation
Well-known techniques for the creation of nanosized surfaces are anodic oxidation of aluminum [26] and wet-chemical deposition of thin layers by means of sol–gel techniques and slurry casting (see Section 7.1). Whilst anodic oxidation is described in detail immediately below, other approaches, referring CVD techniques, will be introduced in Section 7.4.3.

The researchers at TU Chemnitz decided to use anodic oxidation for the generation of a highly porous surface on microstructured aluminum platelets and microchannels, respectively. The experimental protocol was based on performing the oxidation process with aqueous oxalic acid which resulted in a thin layer of Al_2O_3. Such an oxide layer has a very regular pore structure oriented vertically to the flow direction. The depth of the pores can be adjusted precisely by the operating conditions of the electrochemical process.

This porous nanostructure serves as support for embedding catalytically active components [3, 20]. Figure 7-4 shows a diagrammatic representation of the channel microstructure and the oxide nanostructure. The distribution of the pore diameters which are in the

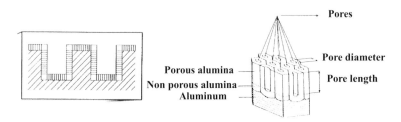

Fig. 7-4. Diagrammatic representation of the cross-section of a microstructured aluminum platelet with an anodically formed oxide layer (left) and nanostructure of the pores inside (right) [18].

range of some tens nanometers is extremely uniform, which has been utilized, for many years, for dyeing of anodized aluminum or for the production of isoporous alumina filtration membranes [27, 28].

By this establishment of porosity, the surface area to volume ratio of the microchannels is increased by a factor of 10,000 to 100,000 and values corresponding to fixed bed catalysts are obtainable. Compared with this benchmark, the pore sizes and depths as well as the distribution of the pores of anodically oxidized aluminium are much more regular.

For testing the performance of a multichannel catalyst structure, the multiple hydrogenation of *cis,trans,trans*-1,5,9-cyclododecatriene (CDT) was chosen using palladium as a catalyst [29]. Starting from this cyclic hydrocarbon with three double bonds, partially and fully hydrogenated products resulted by consecutive reactions. The target product was the thermodynamically unstable intermediate cyclododecene (CDE) [2, 18, 19, 30]. The CDE synthesis has economical relevance by supplying a potential precursor for nylon-12, as an alternative to other existing technically performed pathways.

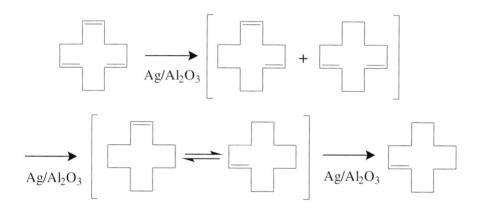

Catalyst Carriers Differing in Pore System, Catalyst Coating, and Packing
A stack of nearly 700 microstructured aluminum foils, manufactured at the Forschungszentrum Karlsruhe (Germany), was used as a microchannel reactor [17]. For comparison, other catalyst carriers, i.e. conventionally coated granules, pieces of activated aluminum wires and fragments of activated aluminum foils, were used. These four types of catalyst carriers differ with respect to selected features, namely their pore system, their distribution of the catalytically active component, and their type of packing, i.e. flow through the reaction zone.

For instance, microchannel platelets are regular with regard to all these features. Due to anodic oxidation, pores of similar size and length are generated which are homogeneously coated with catalyst. As a consequence of the uniform pattern of the precisely microstructured channels, the whole device can be regarded as a regular fixed bed. In contrast, the granules conventionally applied are irregular with respect to all three features, whereas the wire

Catalyst	Type of the catalyst	Pore system		Cross section of catalyst	Distribution of catal. active component		Fixed bed		Cross section of fixed bed
CAT A	conventionally coated granules		+		+		+		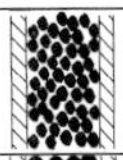
CAT B1	pieces of activated Al-wires	+		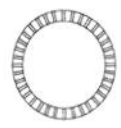	+		+		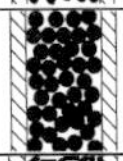
CAT B2	pieces of activated Al-foils	+		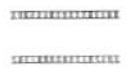	+		+		
CAT C	stack of activated and microstructured Al-foils	+		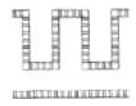	+		+		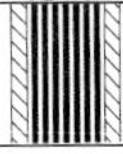

Fig. 7-5. List of catalyst configurations applied and evaluation of their regularity and uniformity with respect to pore size, catalyst distribution, and type of fixed bed [2].

pieces and foil fragments differ from the microreactor only with regard to their irregular packing to a fixed bed (see Figure 7-5).

Performance Improvement by Regular Flow Pattern

In the framework of extensive investigations, the proof of feasibility and process reliability for performing the hydrogenation was initially sought. In this context, the time-on-stream behavior, i.e. the dependence of conversion and selectivity on process time, was characterized (see Figure 7-6). For an 80 h run, a significant decrease in hydrogen conversion was monitored, the CDT conversion being quantitative. After 200 h on-stream, the corresponding data decreased as well. Within 80 h of measurement, the selectivity of the fully hydro-

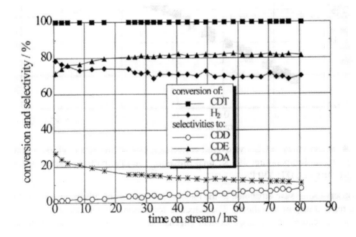

Fig. 7-6. Conversions and selectivities as a function of time on-stream for pieces of catalytically active wires [2].

genated product (CDA) decreased, whereas that of the target product CDE increased. At 100 % conversion of CDT, a selectivity of about 81 % regarding CDE formation was measured.

Another set of experiments focused on the impact of pore length on the selectivity–conversion diagram. For this purpose, two types of pieces of activated aluminum wires, which differed in pore length, were compared. With increasing overall conversion, defined as the sum of the conversion of CDT and the intermediate cyclododecadiene (CDD), the CDE selectivity increased for both pore lengths. However, regarding very high conversions above 90 %, the two types of wire pieces behave differently. For pieces with short pore lengths, the selectivities reached a constant value, whereas selectivities in the long pores decreased.

This result indicates a mass transport limitation causing a too long contact time of CDE with the catalyst. Due to stronger interactions of π-orbitals with the catalyst, CDT and CDD usually displace CDE rapidly from the reaction site, thereby preventing the consecutive hydrogenation to CDA. However, at very low concentrations of these reactants, CDE conversion is more likely. Thus, these investigations show that the surface enlargement achievable by elongation of pores (or decrease in pore size) can only be utilized as long as diffusion is fast enough to follow. In other words, the choice of microchannel dimensions and pore sizes has to be counterbalanced.

A final experiment made is most remarkable since it aimed to demonstrate that the microchannel reactor has clear advantages according to the specifications given in Figure 7-7. The CDE yield was monitored as a function of the overall conversion. Yields of the

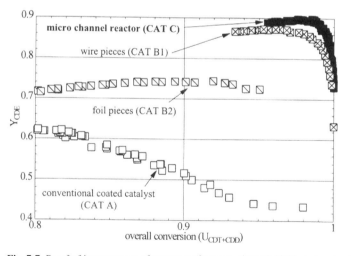

Fig. 7-7. Proof of improvement of reactor performance through highly ordered catalyst configuration and flow conditions. CAT A: pore depth 240 μm, commercial palladium catalyst; CAT B1, B2, C: pore depth 37 μm, anodic oxidized structures of aluminum containing a wet impregnated catalyst, B1 pieces of foils, B2 pieces of wires, C microchannels [2].

conventionally coated granules decrease from about 62 to 44 % within the range of conversions investigated. For the foil fragments, a nearly constant selectivity of about 73 % was found. This underlines the reaction engineering benefits due to a regularly sized pore system and a homogeneous catalyst coating. In addition, the more ordered packing of fixed beds composed of the regular-sized foil fragments reduced the content of dead volumes within the reaction zone. In these dead volumes, the residence time of the product mixture is increased, resulting in enhanced CDE conversion and decreased CDE selectivity.

The best results were achieved for the wire pieces and the microchannel reactor both of which show nearly constant selectivities decreasing at very high conversions. The superior performance of the wires compared with the foils is obviously due to different tight packings since both configurations differ notably in shape, whereas pore sizes and catalyst coatings are equal. Hence performance improvement is achieved, once more, due to a remarkable decrease of dead volumes. Compared with this benefit, the advantage of a uniform flow pattern due to the regularly structured channels seems to be of minor importance, although further increasing reactor efficiency.

Concluding Remarks
In total, the product yield of cyclododecene (CDE) is increased by approximately 45 % at nearly quantitative conversion by using the microreactor compared with standard fixed bed technology. Referring to the best set of operating conditions, the difference in reactor performance still amounts to about 28 %. This clearly outlines potential improvements for consecutive reactions due to regular pore systems with well-defined and relatively short depths of the pores, a homogeneously distributed catalytic component inside the pores, and uniform flow conditions in the microchannels.

7.3.4 H_2/O_2 Reaction

High temperature reactions with millisecond residence times are of increasing interest in catalytic burning and selective oxidation of hydrocarbons. The high rates of these reactions in principle allows one to employ small, light-weight reactors, while maintaining high throughputs. Potential fields of application are energy generation, e.g. in the automotive industry to preheat the combustion catalysts in order to reduce emissions during motor start-up. Another example of use is the selective oxidation of hydrocarbons as performed, e.g., in methane conversion to syngas. A potential driver for miniaturization is the decrease of residence times to the sub-millisecond range which, due to the inherent danger of explosions and flames, is not possible using conventional equipment.

Catalytic Wire Microreactor
In this context, Veser at the Institut für Verfahrenstechnik, Stuttgart (Germany) and researchers of the Institut für Mikrotechnik und Informationstechnik (IMIT), Villingen-Schwenningen (Germany) investigated the hydrogen/oxygen reaction as a model high-temperature process [31]. The microreactor consisted of two stacked silicon chips fabricated

by means of wet chemical etching. The reaction zone is simply defined by a channel, 525 μm deep, about 300 μm wide and 20 mm long, with an embedded commercial platinum wire. The silicon chips are inserted into stainless steel housing, partially manufactured by μ-EDM techniques, thereby mechanically sealing the reaction zone. Heating is achieved by means of resistance heating by electrically connecting the catalyst wire.

This microreactor construction, also conceptually simple, is versatile regarding the variation of catalyst materials and geometric parameters of the micoreactor. Wires, which are available from a vast number of metals and diameters, can be easily exchanged. In addition, they provide a simple means to achieve high reaction temperatures. Potential drawbacks of employing wires as catalysts are the lack of achieving localized heating, non-uniform flow patterns due to less accurate wire positioning, and the potential activity of silicon channel walls.

Safe Operation During Scouting Experiments
As a first remarkable result during experimental investigations, it turned out that the hydrogen/oxygen reaction was not ignited at room temperature, in contrast to findings with conventional fixed-bed technology (compare the results in Section 7.6.1). Ignition was achieved at about 100 °C resulting in a temperature rise to 300 °C. An increase in hydrogen content finally led to a further temperature increase of roughly 1000 °C as evidenced by glowing. The excellent mass and heat transfer properties of the simple wire-based microreactors are demonstrated by the fact that at low flow rates the reaction is restricted to an entrance zone. Only after an increase of flow rate was notable catalyst activity all over the channel length observed. In addition, these first tests revealed high stability of the silicon reactor material. Neither damage of the chips, e.g. crack formation, nor thermally induced irreversible bonding was observed.

$$2\ H_2 + O_2 \xrightarrow[\text{Pt}]{} 2\ H_2O$$

In quantitative measurements, hydrogen and oxygen conversion as well as temperature were monitored as a function of increasing hydrogen content (see Figure 7-8). Dependent on this parameter, three distinct regions characterized by linear and non-linear increases in temperature of the exiting gas were identified. Hydrogen conversion shows a linear increase at low hydrogen contents which flattens and finally becomes complete at high hydrogen contents. This effect, combined with the increase in hydrogen content, results in a parabolic profile of the corresponding oxygen conversion which changes to a linear dependence at high hydrogen contents.

Concluding Remarks
The success of the wire concept is closely related to the constructional simplicity of the corresponding microreactor. Thereby, an elegant way for safely investigating high-temperature reactions is provided which are otherwise difficult to control. It is furthermore noteworthy that, based on the experimental findings, a rough-estimate calculation revealed

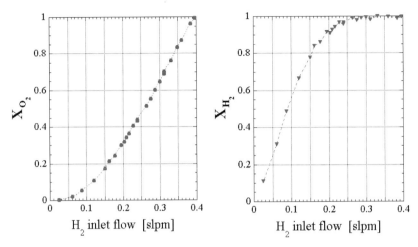

Fig. 7-8. Feasibility of performing a high temperature reaction with millisecond reaction time in a microreactor equipped with a wire, as evidenced for the H_2/O_2 reaction [31]. Oxygen conversion as a function of hydrogen inlet flow (left), Hydrogen conversion as a function of hydrogen inlet flow (right).

that about 100 microchannels are sufficient to heat a standard exhaust catalyst in automobiles within 10 s to process temperature. The same holds for catalytic partial oxidations, rendering microreactors attractive even for production issues. Hence, such simple reactor designs probably may be established as one type of measuring tool for this class of reactions, paving the way for more complex assembled microreactors intended production applications.

7.3.5 Selective Partial Hydrogenation of Benzene

The hydrogenation of the aromatic compound benzene usually proceeds to yield the fully hydrogenated product cyclohexane. Partially hydrogenated intermediates such as cyclohexadiene or cyclohexene are, under normal reaction conditions, so reactive that it is difficult to isolate them in quantitative amounts. However, cyclohexene can be isolated when using special catalysts such as ruthenium and gaseous reaction modifiers such as methanol. In the latter case, adsorption/desorption at the catalytic interface are influenced as well as mass transfer reaction kinetics concerning cyclohexene hydrogenation. Some of these issues are affected by using miniaturized channels as well.

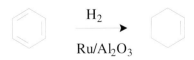

Based on this motivation, researchers at TU Chemnitz (Germany) employed a microchannel reactor [32], developed at the Forschungszentrum Karlsruhe (Germany), which was equipped with 15 aluminum platelets having 450 microchannels of 200 μm x 200 μm x 50 mm size [17]. The platelets were anodically oxidized to enlarge the surface structure and impregenated with ruthenium. Methanol was added as a gaseous reaction modifier.

For a set of operating conditions (60 °C, 110 kPa, 530 ms), a selectivity of 20 % at 13 % conversion was monitored. By adjustment of the methanol content, selectivities could be increased up to 38 %, but at the expense of decreasing the conversion to 5 %. Similar findings were made using zinc as promoter.

7.3.6 Selective Oxidation of 1-Butene to Maleic Anhydride

1-Butene is converted to maleic anhydride via a highly exothermic process using conventional fixed bed technology with V_2O_5 and TiO_2 as catalysts. Maleic anhydride is an important intermediate for alkyd resins and is widely employed in a number of diene syntheses, e.g. Diels–Alder reactions.

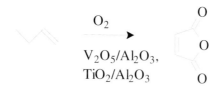

Researchers at TU Chemnitz (Germany) are currently investigating this reaction using a microchannel reactor of dimensions as given in Section 7.3.4 [32], developed at the Forschungszentrum Karlsruhe (Germany) [17]. Preliminary investigations with impregnated anodically oxidized aluminum wires revealed a maleic anhydride selectivity of about 30 % for conversions of 87 %. The construction of a microreactor composed of the same choice of materials is under way.

7.3.7 Selective Oxidation of Ethylene to Ethylene Oxide

A detailed description of conventional processing applied for ethylene oxide generation, disadvantages inherent in this technology and potential for microreactor operation is given in a further section below (see Section 7.5.1).

In order to carry out measurements regarding ethylene oxide synthesis in a microchannel reactor, Kursawe and Hönicke at TU Chemnitz (Germany) applied an evaporation technique to deposit a 200 nm thin catalytically active silver layer [32]. The dimensions of the microchannel reactor made of aluminum, developed at the Forschungszentrum Karlsruhe (Germany), were identical with those listed in Section 7.3.4 [17].

$$\underset{Ag/Al_2O_3}{\xrightarrow{\hspace{1cm}O_2\hspace{1cm}}}$$

An increase in ethylene oxide selectivity with increasing oxygen partial pressure was reported. Consequently, only pure oxygen was employed in further measurements, instead of using an inert gas. The ethylene concentration was set to 4 %. By varying the residence time, conversions ranging from 40 to 66 % were obtained at nearly constant selectivities of about 45 %. This amounts to yields of ethylene oxide of 19 to 29 %. In order to achieve these results, residence times of only a few seconds were needed.

By increasing the ethylene partial pressure to 20 %, selectivity can be further improved up to 55 %, but at the expense of lower conversions. Within the range of residence times investigated, conversions of 15 to 39 % and corresponding yields of 8 to 21 % were found.

7.3.8 Reactions Utilizing Periodic Operation

Periodic operation, i.e. a regular dynamic change in operating conditions, provides a possibility to enhance process performance for selected reactions, e.g. increasing selectivity or enhancing reaction rate [33] (see also [34]). Due to the favorable flow conditions in microchannels, microreactors should be particularly suitable to realize sudden concentration changes of reactants, e.g. rectangular pulses [5], hence, exhibiting a fast response time. In addition, such investigations of periodic behavior would benefit from safe operation in microreactors characterized by small hold-up of flammable or explosive gas mixtures.

In a case study, funded by the German BMBF ministry and accompanied by industrial members, researchers from one institute and two universities are currently evaluating this potential. Research at periodic operation in microreactors is conducted by the University of Erlangen (Germany) regarding isoprene oxidation, the University of Lausanne (Switzerland) regarding alcohol dehydration, and the Institut für Angewandte Chemie, Berlin (Germany) regarding propane dehydrogenation [5]. The microreactor was designed and realized by the Institut für Mikrotechnik Mainz (Germany) on the basis of joint theoretical calculations, assisted in particular by simulations performed at the LSGC–CNRS, Nancy (France) and the Karl-Winnacker-Institut, Frankfurt (Germany).

Residence Time Distribution in Periodically Operated Microreactors
For periodic operation, a narrow residence time distribution is ultimately required. In this context, experimental investigations of the group of Renken at the Swiss Federal Institute of Technology, Lausanne (EPFL) revealed narrow residence time distributions for the fluid flow in a gas phase microreactor designated for fast concentration modulation (see Figure 7-9) [35].

The whole reactor, plates as well as the housing was constructed of stainless steel. Each plate contained 34 channels of 300 µm width, 240 µm depth and 20 mm length. Plates with catalyst coated channels and heat exchanger plates for heat removal are arranged alterna-

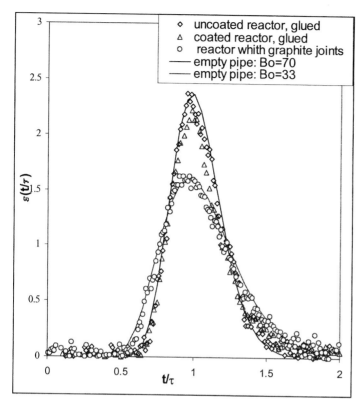

Fig. 7-9. Residence time distributions of a gas flow in a periodically operated microreactor using uncoated glued or catalyst-coated glued microchannels. In addition, the corresponding plot for a graphite-sealed microreactor is given [35].

tively. The reactor volume is ca. 0.9 cm^3. The residence time distribution was measured at flow rates of 0.2–0.3 ml/s.

The experimentally determined Bodenstein number

$$B_0 = \frac{u \cdot L}{D_{ax}} \tag{7.1}$$

u: velocity; L: length; D_{ax}: axial diffusion coefficient.

was found to be 70 indicating that the microreactor behaved almost like a plug-flow reactor. In addition, it was found that small details concerning the manufacturing process, as changes in the sealing method chosen, strongly affected the width of the residence time distribution. In contrast, the deposition of a catalyst layer had nearly no influence on the hydrodynamic behaviour.

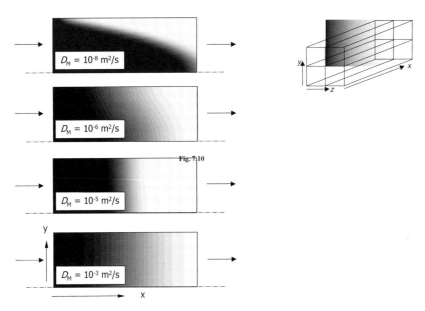

Fig. 7-10. Concentration profiles after a step function at the inlet. Simulations performed by Schreiber and Weber at the Institute of Fluid Dynamics (Chair: F. Durst) at the University of Erlangen [36].

Further fluid dynamic investigations, based on CFD simulations, were carried out by Liauw and Walter in cooperation with Weber and Durst at the Universität Erlangen (Germany) [36]. The simulated response to a step function in microchannels of 71 to 1200 µm width and depth and 5 to 40 mm length was determined at linear velocities of 0.03 to 2 m/s. The simulation results illustrate the major impact that radial and axial diffusion can have on the concentration profile established in microchannels, dependent on the diffusion constant D. In the case of small values, e.g. $D = 1 \times 10^{-8}$ m^2/s, the concentration profile follows the parabolic flow profile (see Figure 7-10). In contrast, the concentration step function is significantly influenced by axial diffusion at high values, e.g. $D = 1 \times 10^{-3}$ m^2/s. Consequently, both effects lead to a considerable broadening of the respective residence time distributions. However, at medium values of the diffusion constant, e.g. $D = 1 \times 10^{-5}$ m^2/s, radial diffusion counteracts the parabolic flow profile, thereby leading to a narrow residence time distribution. Diffusion constants of this magnitude are typical for gases, thereby fulfilling one basic requirement for performing periodically modulated gas phase reactions in microreactors.

Isoprene Oxidation to Citraconic Anhydride
For isoprene oxidation to citraconic anhydride, it is envisaged to increase selectivity by means of periodic operation, similar to findings already gathered for butane oxidation [37]. The key for reaching this goal is to employ a more selective reactant species as achievable

Fig. 7-11 Photographs of an assembled and disassembled microchannel reactor for periodic operation [5, 38].

in a non-periodic process. By repeated concentration cycles of oxygen and isoprene, a distinct oxygen species which selectively reacts with isoprene is reversibly generated at the catalyst surface. In addition, periodic operation prevents contact of isoprene with unselective gaseous oxygen. This change in reactants affects the reaction pathway, hence potentially favoring partial oxidation to citraconic anhydride over total oxidation of isoprene.

$$\text{isoprene} \xrightarrow[\text{V}_2\text{O}_5/\text{Al}_2\text{O}_3]{\text{O}_2} \text{citraconic anhydride}$$

As catalyst supports, anodically oxidized aluminum platelets were used in order to enlarge the support surface area (see Figure 7-11 and also Section 7.3.2). These platelets are characterized by two flow distribution regions connected to bore-holes for reactant feed and product withdrawal. A microchannel array is inserted between these distribution regions. This type of platelet configuration was used for the two other reactions as well, the characteristic internal dimensions being adapted to the respective process. V_2O_5 was employed as a catalytically active component by means of wet chemical deposition.

The influence of the geometry of the array and the other fluidic structures on flow distribution and shape of concentration pulses is currently being investigated. For instance, calculations by Matlosz and coworkers at LSGC-CNRS (Nancy, F) showed how different inlet geometries concerning the bore-holes, referred to as U- and S-type, result in different distributions of single flows in the microchannels [39].

Catalytic Dehydration of Alcohols

In the case of the catalytic dehydration of alcohols a considerable increase in reaction rate can be achieved by periodically switching the reactant concentration to zero, referred to the stop effect [40]. It is known that this phenomenon is related to adsorption/desorption processes which were theoretically described by two different models. Thus, the benefits of studying alcohol dehydration in periodically operated microreactors are two-fold: to evaluate their potential as production devices, e.g., by increasing the space–time yield, or to utilize them as measuring tools, e.g., determining the validity of a reaction model.

$$\searrow\!\!-OH \xrightarrow{\gamma\text{-Al}_2O_3} \diagdown\diagdown$$

A suitable catalyst carrier system for alcohol dehydration is γ-alumina deposited on stainless steel platelets. Calculations predicted the use of 10 to 20 μm thick coatings of a specific surface area of about 25 to 50 m^2/g. These relatively thick layers were achieved by a wash coating technique using γ-alumina powder and a boehmite suspension, yielding an agglomerate of particles connected by a binder. By means of a masking process, selective coating within the microchannels could be achieved.

Propane Dehydrogenation
In the case of propane dehydrogenation, an exothermic oxidative process is coupled to an endothermic non-oxidative process, both based on propane as reactant and propene as product [41]. However, the non-oxidative process is accompanied by coke formation. In order to burn off this deposit, periodic operation alternately feeding propene/oxygen and air is required.

$$\diagdown\diagdown \xrightarrow[\text{V}_2\text{O}_5]{(\text{O}_2)} \diagdown\diagup$$

Stainless steel could not be employed as construction material due to undesired catalytic activity. Therefore, titanium was chosen as the material for the reaction platelets. By means of plasma-assisted sputtering and chemical vapor deposition, catalyst layers were generated. Vanadium was applied in the oxidative zone, whereas platinum was chosen as catalyst for the non-oxidative reaction.

Concluding Remarks
It has to be emphasized that the investigation of periodic operation in microreactors has started just recently, in contrast to most of the other reactions referred to here, being entirely completed. So far, results have not been published. However, the activities already revealed important general information on the design of reactor/heat exchanger platelets, in particular concerning flow equipartition. Regarding microfabrication, different interconnection techniques, e.g. glueing, flat sealing, or diffusion bonding, were evaluated and optimized with respect to different construction materials. First promising approaches for catalyst coating in microchannels were successfully implemented.

7.4 Microsystems with Integrated Catalyst Structures and Heat Exchanger

7.4.1 Oxidative Dehydrogenation of Alcohols

Limitations of Current Scale-Up Procedures and Potential of Microreactors Therein
In the chemical industry, existing methodologies regarding scale-up procedures, i.e. to transfer results from laboratory to a production scale, still have a large potential for further optimization. It is economically and ecologically rational to gain information about kinetics and hydrodynamics already at the stage of laboratory reactors. In one reference from industry [42], a schematic outline of a possible strategy for this purpose is given. Kinetics of a large-scale process can be determined from measurements of laboratory-reactor performance, if calculations of fluid dynamics of a large-scale reactor are available. Based on these data, conclusions regarding the performance of the large-scale reactor can be drawn.

However, using conventional laboratory-scale equipment, it often turns out to be difficult to scale-up a process in one step. In many cases further measurements in pilot plants exceeding the size of the laboratory-equipment but still smaller than the production plant are needed. The reason for this lack of reliability of data gained in the laboratory is due, e.g., to considerable changes of fluid dynamics when scaling-up by more than a factor of ten.

Thus, the specific designs of large-scale reactors were often found by trial-and-error development in the past [23]. In some cases, designs were chosen on the basis of analogy considerations due to existing knowledge about a "supposed-to-be-similar" process. This procedure has two distinct disadvantages. First, the development is costly and time consuming. Large costs may arise needlessly if a large-scale reactor is operated over years at non-optimum performance, since the industrial research conducted to optimize the process adds a similar amount to the overall costs.

In addition, it takes years until an existing reactor design may be changed, thus hindering the realization of innovative approaches. In many cases, it is even not known if the existing reactor is close to or far from optimum reaction conditions. Hence, process development is conducted without exactly knowing the scope of potential benefits. This lack of a reliable measure for the ultimate potential of a reaction is referred to as a second drawback of existing laboratory-technology in the field of process scale-up.

Microreactors can help to solve this situation. Due to their small linear dimensions, transport processes can be enhanced so that the "underlying" chemistry can be analyzed without any interfering effects. Under some circumstances, a nearly ideal reactor behavior may be realized, being characterized by spatially independent values of concentration and temperature and a well-defined residence time [43, 44].

Moreover, important parameters for fluid dynamic characterization, such as interfacial areas in multiphase reactions, are either readily available or even predetermined by the reactor geometry. An example for the latter type is the formation of falling films in a microreactor, showing a close resemblence between theoretical data based on calculations from film theory and experimental values (see Sections 8.1 and 8.3).

Use of a Microchannel Reactor for Process Optimization

Researchers at BASF, Ludwigshafen (Germany) were among the first to investigate the potential of microreactors for industrial process development [42]. They used a micro heat exchanger/reactor from the Forschungszentrum Karlsruhe (Germany) [45] for analysis of gas phase reactions. Platelets of the micro heat exchanger applied contained reaction channels of a depth of 300 μm, a width of 400 μm and a length of 10 mm. For this reactor geometry, a total heat transfer coefficient of 20,000 W/m^2 K was reported [17], exceeding conventional heat exchangers by a factor of at least 20.

Already the very first investigations were successfully able to demonstrate the large potential of microreactors for process development. For a process carried industrially out in a tube-bundle reactor, complete conversion and an increase of the selectivity by 10 % were achieved using this microreactor [42]. Unfortunately, no details concerning the reaction were given.

Dehydrogenation of a Derivatized Alcohol

Much more detailed information was available in a series of publications and oral contributions at several conferences regarding an oxidative dehydrogenation process, i.e. the conversion of an alcohol to an aldehyde [23]. This process is one example among others within a homologous series of reactions which differ in the respective alcohol reactant regarding its substituent, replacing the hydrogen at the functional carbonyl group. The simplest molecule of this series results if the substituent is hydrogen, i.e. referring to formaldehyde. The corresponding gas phase process was developed at BASF about 100 years ago using oxygen as reactant and a silver catalyst [23, 42].

$$CH_3OH \xrightarrow[Ag]{O_2} H\text{-}CHO$$

At this time, no deep knowledge about kinetics or fluid dynamics could be used as a design criterion for construction of the large-scale reactor. Nevertheless, the pioneering researchers managed to establish a profitable process in a large pan-shaped reactor of a diameter of 7 m [23], referred to as a short-path reactor [42]. A 2 cm layer of silver grains on an inert porous support served as catalyst. At almost quantitative conversion, a selectivity of 95 % was achieved.

Stimulated by these results, BASF tried to utilize the pan-shaped reactor design for a similar process several years ago, namely the synthesis of a derivatized aldehyde. This product, which contains an olefin function as substituent, is an intermediate in a multi-step synthesis. The feasibility of the concept was tested by determining the performance of a small laboratory reactor having a pan of 5 cm diameter.

$$R\text{-}CH_2\text{-}OH \xrightarrow[Ag]{O_2} R\text{-}CHO$$

Use of a Large-Scale Pan Reactor

A selectivity of 85 % was obtained at a conversion rate of 50 %. Since this was regarded as being economically attractive, a large production reactor with a pan of 3 m diameter was built. It must have been disappointing for the optimistic BASF researchers that the selectivity of this reactor dropped to 45 %. Although this was nearly unacceptable, the process had to be run over years with this suboptimal reactor performance, thereby causing high costs.

From the beginning of the search for an alternative concept, it was clear that heat removal plays an important role in process performance. Calculations showed that about 50 % of the reaction heat was dissipated through the vessel walls in case of the laboratory-scale pan reactor. Due to the smaller surface-to-volume ratio, the large-scale reactor, instead, was operated completely adiabatically. This resulted in a temperature rise of about 150 °C compared to 60 °C for the laboratory reactor. Consequently, the dramatic drop in selectivity can be attributed to side and follow-up reactions of the thermally labile alcohol reactant and the aldehyde product. In the case of formaldehyde synthesis, the same scenario of reactor scale-up did not affect the selectivity since all species involved in the reaction were thermally stable.

Use of a Large-Scale Multitubular Reactor

Knowing that heat removal is a key issue, reactor concepts specifically developed for highly exothermic reactions were evaluated for their suitability for the current problem. As one candidate, a multitubular reactor, also referred to as a tube bundle reactor, was equipped with bundles of 14 mm diameter, corresponding to the smallest size commercially available. Usually, these reactors are at least several meters long in order to avoid hot-spot formation as a consequence of insufficient backmixing [42]. The increase in residence time allows nearly isothermal operation as evidenced for, e.g., ethylene oxide synthesis. However, it was also known that a longer contact time at high temperatures may lead to an increased contribution of thermal side reactions.

For this reason, a short multi-tubular reactor having tubes with 200 mm length was finally chosen for the synthesis of the derivatized aldehyde. This reactor concept combines short residence times with efficient heat transfer (see Figure 7-12).

Use of a Microchannel Reactor

The hot-spot in the tubes of this reactor was limited to about 60 °C, comparable to the laboratory reactor. As a consequence, a selectivity of 85 % was achieved on a production scale. The logical further development of the multi-tubular reactor is the use of a microchannel reactor, as described above [42]. This microreactor allows, even for conversions higher than 50 %, operation with isothermal conditions and residence times an order of magnitude lower. After a few experiments, a selectivity of 96 % was obtained at conversions higher than for the multi-tubular reactor (see Figure 7-13). Therefore, a selectivity increase of 16 % with respect to the production reactor was achieved, demonstrating the large potential for further process improvement.

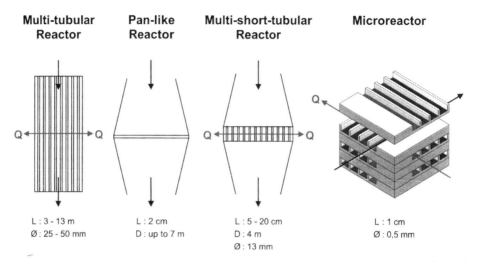

Fig. 7-12. Reactor configurations for gas phase synthesis of aldehydes by oxidative dehydrogenation of alcohols [44].

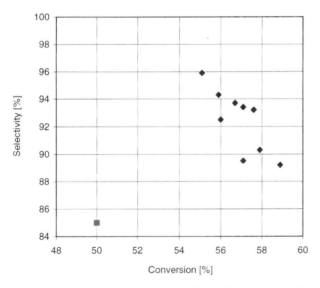

Fig. 7-13. Comparison of the performance for aldehyde synthesis of a microchannel reactor (rhombus) and a short multi-tubular production reactor (square) by means of a selectivity–conversion plot [44].

This was a promising result, also from a commercial point of view, since the reactants are utilized more effectively and a decrease of the amount of side products reduces the expenditure for subsequent separation and purification.

Concluding Remarks
BASF researchers emphasized that such developments cannot be performed using conventional laboratory techniques, but are only feasible by use of microreactors. They expect a significant reduction of development periods and considerable process improvements by using microreactors as a versatile development tool for optimizing production scale synthesis.

7.4.2 Synthesis of Methyl Isocyanate and Various Other Hazardous Gases

Pioneering work on the implementation of microreaction technology in the chemical industry was made by internal investigations at DuPont, Wilmington (USA) [43]. In the framework of several studies, each being fixed to one chemical process, the technical feasibility and potential of microreaction systems was analyzed. The development work was accompanied by detailed analysis of the advantages and application fields of microreactors.

The main issues of the work conducted relied on the substitution of a large plant by many distributed microreaction facilities. Therefore, mobile reaction units had to be developed allowing production at the point-of-use and an adjustment of production capacity to varying needs. Although their reactor performance is somewhat lower than that of the plant and, thus, the overall economic balance may not be superior, the substitution may be profitable for the following reasons. A localized use of microreactors eliminates risks and costs concerning transport and storage and also minimizes safety expenditure due to a smaller total hold-up of hazardous chemicals.

Design Issues for a Modular Microreactor
In the studies, reactions (see Table 7-1) were carried out in modularly constructed devices consisting of heat exchangers, catalyst chambers and other components not listed explicitly. These microreaction systems comprised a stack architecture of bonded microfabricated silicon wafers, each carrying one reactor component. Only a sketch of the construction of the microreactor was given within the main contribution [43] and the design of the components was not described in detail. A patent filed by DuPont gives a deeper insight into the design strategy chosen [46]. In this context, a patent of the Academy of Science of the former GDR on microreactor construction, filed prior to the DuPont patent, is noteworthy [47]. Ten years before research concerning microreaction technology started, basic ideas for construction were outlined in this patent.

The main issues for design development were to achieve a proper manifolding of fluid streams in a large number of parallel channels and efficient heat management. A threefold hierarchy for flow partitioning was developed, starting from parallel operated reaction units which comprise many individual wafer units, finally containing a multitude of reaction channels.

It was pointed out that wafer-based microfabrication technologies are, in particular, suitable for accurate flow partitioning due to their high structural precision and precise alignment capabilities [43]. As a suitable manifolding principle, the use of small orifices with precise hole diameters at the entry and exit of each channel was recommended.

Heat transfer was accomplished between two adjacent wafers, one carrying the reaction site, the other being dedicated to an energy source or sink. Concerning heat removal in exothermic reactions, the use of circulating coolants was suggested. As one example of heat supply in the case of endothermic reactions, resistance heating, e.g. using ceramic plates, was proposed. Thereby, it was aimed to minimize temperature gradients along the flow axis and between single channels. Facing economic targets, heat manangement should be improved by recovering energy, e.g. use of an excess of heat released for preheating purposes.

Experimental Case Study Based on Different Types of Reactions
In order to evaluate these potential benefits of microreactors, a broad experimental program based on seven reactions was performed (see Table 7-1). The spectrum ranged from hydrogen cyanide formation to the synthesis of chlorinated silanes. This set of reactions can be divided into four classes, when characterizing the processes with respect to their temperature, hazards, activation (catalysis) and initiation (photochemical). Each reaction required a unique assembly of different components. For instance, the reaction chamber had to be equipped with a transparent plate to permit photochemical processes.

Table 7-1. Selected reactions tested in microfabricated silicon devices.

Reactants	Main products	Reaction category
Cyclohexylamine, phosgene	Cyclohexyl isocyanate	High temperature, hazardous
Butylamine, phosgene	Butyl isocyanate	High temperature, hazardous
Hydrogen chloride, oxygen	Chlorine, water	Catalytic, hazardous
Carbon monoxide, chlorine	Phosgene	Catalytic, hazardous
Methylformamide, oxygen	Methyl isocyanate	Catalytic, high temperature, hazardous
Methane, ammonia	Hydrogen cyanide	Catalytic, high temperature, hazardous
Dichloromethylsilane, chlorine	Trichloromethylsilane, hydrogen chloride	Photochemical, hazardous

Synthesis of Methyl Isocyanate
As a case study, the formation of methyl isocyanate from methylformamide and oxygen was carried out. This reaction refers to the class of high-temperature, hazardous, catalytic reactions which need no photochemical initiation. Moreover, the high exothermicity demands intense cooling.

$$CH_3\text{-}NH\text{-}CHO \xrightarrow[\text{Ag}]{O_2} CH_3\text{-}N\text{=}C\text{=}O$$

The reaction was carried out in a microreactor consisting of three etched silicon wafers carrying microstructured components and two capping wafers (see Figure 7-14). Reactants

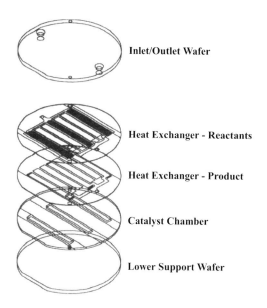

Inlet/Outlet Wafer

Heat Exchanger - Reactants

Heat Exchanger - Product

Catalyst Chamber

Lower Support Wafer

Fig. 7-14. Exploded view of the DuPont microreaction system composed of silicon wafers [43].

were mixed before entering a micro heat exchanger channel packed with polycrystalline silver particles. Analysis of the product stream was performed by means of capillary gas chromatography.

Conversions up to 95 % for the reactant methylformamide were obtained for operating conditions similar to those of a standard laboratory-scale reactor system. However, the selectivity determined was lower. The chemical and thermal compatibility of the construction material silicon turned out to be sufficient, showing no corrosion or channel coating with soot layers or plugging with particles.

The desired uniform temperature profiles could not be achieved, i.e. an operation between isothermal and adiabatic was established. As a consequence, a relatively high temperature maximum of up to 300 °C was found.

Concluding Remarks

The experiments successfully demonstrated the feasibility of a microreactor for catalytic processing of hazardous materials at high reaction temperatures. In this context, the modular reaction device of DuPont appears to be a promising approach towards production at the point-of-use, e.g. for laboratory purposes.

As an outlook, the DuPont researchers provide a practical solution for process control of such systems. Since it is not feasible at present to measure certain physical properties at each channel among a large number of channels, the authors propose to use simplified concepts. Referring to flow control, for instance, flow rates should be measured only at the point of the manifolding zones building up a defined pressure. An attainment of equal process conditions within the single channels is guaranteed by replication of identical shapes.

7.4.3 H_2/O_2 Reaction in the Explosion Regime

Exploitation of New Process Regimes
For a number of reactions the use of a range of operating conditions in large-scale reactors is prohibited for safety reasons. This holds, in particular, for processes carried out with oxygen as one reactant, frequently characterized by the presence of extended explosive regimes. This limited access to part of possible process regimes, e.g. with respect to temperature and pressure, may restrict the efficiency of a process. For instance, a possible operation at higher process temperatures or pressures could increase conversion. Hence, higher space–time yields would result, in the case of using a reactor capable of safe operation, in potentially explosive regimes.

Generally, explosions are due to an uncontrolled increase of reaction rate during operation of two-fold origin. Thermal explosions are caused by the release of large amounts of heat, stimulating further reaction and, consequently, again increasing heat release. In case of insufficient heat removal in the reactor, thermal runaway builts up, finally leading to explosion. Another type of explosion, referred to as isothermal type, is characterized by a steadily rising number of reaction chains, as often found for radical reactions. The key to control this situation lies in the collision of some radicals with a solid material, e.g. the reactor material, in order to deactivate these reactive species.

Both pathways to impede explosion, due to either better thermal control or wall deactivation, are supposed to be more efficient on reducing characteristic scales of the reactor. A decrease of linear dimensions, e.g. the diameter of a stirred tank, facilitates heat transport from a reacting fluid to a cooling medium. A larger surface to volume ratio results in an increased probability of collision of gas molecules with a wall which is used in flame arrestors.

Hence past experience has shown that the most severe explosions usually occur in large reactors, while the extent of possible damage can be reduced or even totally avoided when using laboratory scale equipment. A typical example thereof is direct fluorination with elemental fluorine (see Section 8.2.5) which, despite promising results and relatively safe operation in small stirred flasks, cannot be scaled up to a large industrial process. A further miniaturization of laboratory equipment ultimately leads to the use of microreactors. These devices benefit from a small fluid layer thickness and large interfacial areas, e.g., yielding high heat transfer coefficients and channel diameters below the average distance between two molecule collisions.

Although processing in explosive regimes seems to be an ideal approach for microreactors, already being predicted in early publications [42, 43, 48], some years passed until the first experimental work therein was reported. Most likely this was hindered by both the relatively extensive experimental expenditure and a missing industrial driver, being more focused on avoiding explosions than on practically exploiting the corresponding regimes. In this context, first attempts were aimed at the demonstration of the feasibility of safe operation, using well-known explosive reactions such as the H_2/O_2 reaction to water.

H_2/O_2 Reaction in Cross-Flow Microchannel Reactor
The group of Schüth at the Max-Planck-Institut für Kohleforschung in Mülheim (Germany) carried out basic research concerning the H_2/O_2 reaction [24] using a heat exchanger/reactor of the Forschungszentrum Karlsruhe [17]. This process served as a model reaction, compared with other processes simplified by the absence of side or follow-up reactions. A transfer of the results of this model process, especially concerning heat removal, to industrially important processes was envisaged. In this context, the authors indicated that results may be transferable to other oxidation reactions and mentioned explicitly the use of the microreactor as part of a fuel cell system.

$$2\,H_2 + O_2 \xrightarrow[\mathrm{Pt/Al_2O_3}]{} 2\,H_2O$$

In order to yield a cross-flow system, two types of stainless steel platelets were alternately stacked. The reactant flow was guided through channels of 100 µm x 200 µm cross-section, while smaller heat exchange channels (70 µm x 100 µm) carried nitrogen as a cooling fluid. For reasons of passivation against corrosion, the reaction platelets were coated with an alumina layer by means of a chemical vapor deposition (CVD) process at atmospheric pressure (see Figure 7-15).

As an organometallic precursor for the alumina layer, aluminum isopropyl oxide diluted in nitrogen as carrier gas was fed into the hot microreactor, thereby inducing decomposition. No blocking of the reaction channels due to uncontrolled alumina deposition was observed. Platinum as catalytically active species was deposited on the alumina layer by a variant of wet chemical impregnation. This process, referred to as incipient wetness, is

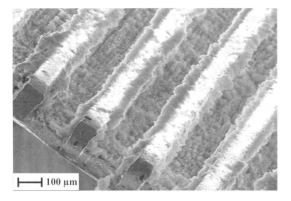

100 µm

Fig. 7-15. Scanning electron micrograph of an alumina layer as catalyst support deposited by means of a modified CVD process (by courtesy of Schüth, MPI Mühlheim).

especially suited for a precise setting of the catalyst concentration. Starting from repeated wetting with an aqueous solution of a platinum precursor, the final state of the catalyst was realized by standard heating and reduction processes.

Experimental Set-up and Protocol
Although it stands to reason that operation inside the microreactor should be inherently safe, this does not hold for the external fluidic peripherals before and after. A one-way valve was introduced to prevent potential explosion from gas entering the gas mixing and supply units. A volume reduction of the exit tubes served for a restriction of combustion of the product gas mixture after leaving the microreactor.

The microreactor was packed between two aluminium blocks equipped with heating cartridges. This assembly was thermally insulated by embedding in a ceramic frame. Heat generated by the heating cartridges was only used to initiate the reaction, followed by further temperature increase due to reaction enthalpy. The yield of the reaction product was determined by gravimetric control of water absorption within a dehydrated molecular sieve.

Typical flow rates of the nitrogen carrier gas amounted to 1.0 l/min; the nitrogen coolant flow was about 3 to 10 times higher. The oxygen flow rate was kept at either 0.1 or 0.2 l/min and the hydrogen flow was either set equal or added according to the stoichiometric ratio. Thus, the volume content of the reactants was approximately 38 %. After this value was reached by stepwise increases of the hydrogen flow rate, the temperature of the leaving gas mixture steadily increased from 80 °C as a function of operation time from initiation of reaction.

Control of Operating Temperature by Flow Variation
The first experiments were aimed at control of the operating temperature by variation of the flow rates of both reactants. With respect to the oxygen and hydrogen rates mentioned

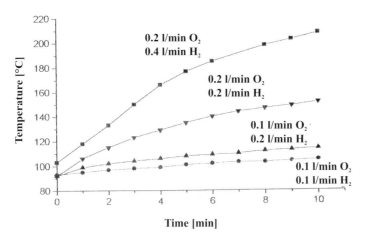

Fig. 7-16. Control of operating temperature as a function of residence time and reactant flow rates for the H_2/O_2 reaction using a cross-flow heat exchanger/reactor [43].

above, four experimental runs were carried out (see Figure 7-16). For the experiment using the lower oxygen and hydrogen concentrations, a temperature increase of only 10 °C within 10 min operation was reported. Instead, the temperature rose about 110 °C in the same time interval for the higher flow rates. The experiments with stoichiometric ratios resulted in a larger temperature increase compared with those applying equal rates of oxygen and hydrogen.

For one set of operating conditions (0.2 l/min O_2, 0.4 l/min H_2, 1.0 l/min N_2 diluent, 0.4 l/min N_2 coolant) the conversion of the reactants to water was determined to be 100 %. The rate of water production amounted to 0.33 g/min. Higher reaction rates may be achievable by applying still higher contents of hydrogen in the nitrogen carrier. In order to restrict the corresponding temperature rise within practical limits, the gaseous coolant might be replaced by more efficient liquid coolants, although this was not explicitly shown in the study.

Concluding Remarks

On the basis of these results, the experiments successfully demonstrated the feasibility of safe operation of a potentially explosive reaction in a microreactor. The contact time for the O_2/H_2 reaction was sufficient for high conversion. This result indicated that, by proper design of the microreactor, the composition of the exit gas can be kept completely out of the range of an explosive regime. Since the potential danger resulting from combustion in the feed lines in front of the microreactor can be minimized by integration of a micromixer, a safe process performance is given for the total fluidic line from gas supply to product separation and analysis.

Another important result of this study refers to the large temperature increase of the reactant gas. Although an external heat supply and an internal coolant were used, the reaction temperature was significantly dependent on the operating conditions. In the framework of the study, this was desired, but it may be detrimental in other cases due to a too high temperature increase. This clearly shows the limits of gas/gas micro heat exchangers concerning highly exothermic reactions. A characterization of the axial temperature profile inside these reaction and heat exchange channels would yield valuable information for further microreactor design. Practical solutions to overcome this problem rely on the use of liquid coolants, as mentioned above, or modifications of microreactor geometry and construction materials.

7.5 Microsystems with Integrated Catalyst Structures and Mixer

7.5.1 Synthesis of Ethylene Oxide

Comprehensive Approach for Microreactor Component Realization

The oxidation of ethylene to ethylene oxide using a silver catalyst in a microreactor was investigated with cooperation between researchers at the Institut für Mikrotechnik Mainz (Germany) and the group of Schüth at the Max-Planck-Institut für Kohleforschung, Mülheim

(Germany) [1, 49–51]. The former group was responsible for theoretical calculations, design, and construction of the microreactor, whilst the latter group performed testing and performance evaluation.

$$ \underline{\quad} \quad \xrightarrow[\text{Ag}]{\text{O}_2} \quad \overset{\text{O}}{\triangle} $$

The work was embedded in the framework of a long-term research project, founded by the German government and conducted by a project team including seven research institutes and an advising committee with members from four chemical companies. Therefore, it was possible to establish several experimental and theoretical research branches of strong mutual interaction. Hence the time schedule allowed the research issues to be extended from the development of practicable fabrication strategies for gas phase reactors towards their detailed experimental characterization completed by extensive work on process simulation.

Although the work was oriented towards fundamentals for the most part, some results already identified and evaluated potential applications by comparison with an industrial benchmark. For this reason, the microreactor performance was referred to the large-scale ethylene oxide formation process which currently is performed by direct oxidation with air or pure oxygen over alumina-supported silver catalysts. This reaction is performed with catalyst supports having pore sizes between 0.5 and 50 μm and at temperatures ranging from 250 to 300 °C. In addition, kinetic data for the reaction are well known, allowing a further point of comparison.

Ethylene Oxide Synthesis
Since the specific adaptation of existing microfabrication technologies to the needs of microreactors was still in its infancy at the start of the project, it was only possible to a certain extent to re-establish completely the industrial process in the microreactor. In particular, this holds for the catalyst system, the complex supported catalyst carrier being simply replaced by polycrystalline bulk silver material. In addition, promoters such as chlorinated hydrocarbons were not added to the feed gas mixture. For this reason, it was expected that the efficiency of the microstructured bulk catalyst material, in terms of conversion and selectivity, would be lower than the benchmark. Thus, the magnitude of this gap served as the real measure of the microreactor performance.

The partial oxidation of ethylene to ethylene oxide is well suited for demonstrating the advantages of microreaction technology for the following reasons. Similar to the dehydrogenation of cyclododecatriene [29], this reaction is part of a series of reaction steps that can potentially occur. Whilst control over residence time and mass transfer was the key in the latter application, ethylene oxide formation mainly benefits from processing under isothermal conditions, thereby avoiding any hot-spot generation.

Under the reaction conditions chosen, the target ethylene oxide is liable to undergo further oxidation to carbon dioxide. Moreover, the reactant ethylene may also convert to this product, referred to as total oxidation. These two reaction pathways, yielding the thermodynamically stable product carbon dioxide, will benefit from any increase in temperature. Even worse, they may contribute to a further temperature rise, due to the large difference in reaction enthalpy compared to the partial oxidation. Thus, temperature control is inevitably needed to reach an acceptable level of ethylene oxide selectivity.

Mixer/Catalyst Carrier Microreactor with Stacked-Plate Architecture

To exert efficient thermal control, a microreaction system was realized, including components for mixing and catalysis. Due to the high thermal conductivity of the bulk catalyst material and the small reaction volume, internal heat transfer, e.g. by employing microstructured cooling channels, was not needed. Calculations of the temperature profiles within an externally heated microreactor confirmed a maximum temperature deviation of 1 K due to reaction, even at high reactant conversions. As a practical solution, a forced convection flow oven was used to reach the temperatures needed, ranging from about 200 to 360 °C. The microreactor system can be connected to miniaturized gas separation and GC analysis components, developed in the framework of the above-mentioned project as well.

In Figure 7-17 the disassembled microreaction system is shown, comprising a central housing with chambers as containments for stacks of mixing and catalytic platelets. The reaction gases ethylene and oxygen enter the stack of mixing platelets via feed tubes, welded to the housing from opposite sides, are guided from there to a diffusion volume followed by a multichannel passage through the catalyst stack. Due to a proper choice of materials, e.g. stainless steel alloy, silver and nickel, and adequate sealing techniques, e.g. mechanical compression by means of graphite foils, the ethylene oxide microreactor was designed to withstand temperatures up to 360 °C and pressures up to 25 bar.

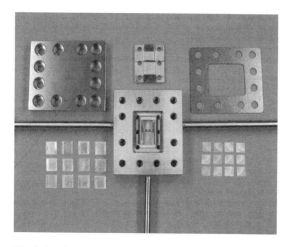

Fig. 7-17. Disassembled microreaction system for ethylene oxide synthesis with central housing, feed tubes and platelets for stacking catalyst and mixing structures [50].

Fabrication of the reactor housing was performed by conventional precision engineering in combination with micro electro discharge machining and welding [50, 51].

Microstructured platelets for mixing and catalysis were realized by a combination of photoablation via excimer laser radiation and electroforming (Laser-LIGA). In the case of the mixing platelets, micromilling techniques were additionally applied in order to manufacture the residual part of the platelet excluding the curved channel system.

Micromixer with Curved Channels

All platelets of the micromixer are characterized by a curved channel system of varying channel width and length, as discussed in detail in Section 3.7.3, where the inlets of the channels of a platelet are alternately connected either with the ethylene or oxygen feed side. The special channel geometry, determined by means of fluidic simulation, guarantees for flow equipartition at the channel outlets by adjusting the pressure drop.

The turn of the gas flow through a right angle allows the precise and homogeneous formation of a multilaminated layer system entering the diffusion volume as a multi-entrance flow. The diffusional path of the gas molecules between adjacent layers is, for equal reactant flows, determined by the platelet thickness of about 200 μm. Calculations proved that a length of the mixing zone of 1 mm is sufficient to guarantee complete mixing gases.

In Figure 3-34 a schematic of the stack of platelets and the multilamination process as well as an SEM image of the assembled micromixing platelets was already shown. The width of the channels on the platelets ranges from about 150 to 470 μm, the depth being 50 μm. The platelets were made of nickel coated with gold to prevent catalytic activity.

Catalyst Zone with Shallow Channels

Behind the diffusion volume, the catalyst zone comprises a stack of solid silver platelets with parallel straight microchannels. The channels have a width of 500 μm and a depth of 50 μm to realize short diffusion paths to the catalytic surface, the length being set to 9 mm.

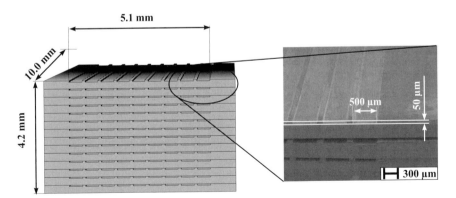

Fig. 7-18. Schematic of a stack of silver catalyst platelets with parallel shallow channels (left) and scanning electron micrograph of such a stack forming a multichannel catalyst zone [51].

Figure 7-18 shows a schematic of the stack comprising the catalytic platelets and an SEM image of the real microstructures.

Catalyst Activation

Before carrying out the reaction, the silver platelets were exposed to a combined oxidation and reduction cycle, referred to as OAOR process [1], in order to increase the specific surface area of the catalyst. Although the increase in surface area could not be precisely judged, an analysis of chemisorption data suggests an enhancement of a factor of about two to three.

Using new silver platelets without any surface area increase by oxidation, no conversion was observed. Even after utilizing the OAOR process, only a small catalytic activity could be monitored. Only a long-term thermal treatment, 100 h heating at reaction temperatures, was finally successful, most likely due to diffusion induced molecular rearrangements of the catalytic surface structure. This was demonstrated by reaching a reaction rate of 1.5×10^{-4} mol s^{-1} m^{-2}, comparable to values reported in the literature for silver powder [52].

Experimental Investigations, Partially Performed in Explosive Regime

During the course of the experimental studies, ethylene and oxygen were diluted by a nitrogen carrier gas to yield concentrations ranging from 1 to 6 vol.-% and 5 to 50 vol.-%, respectively. The total gas flow amounted to 1 to 5 l/h and the pressure was varied up to 20 bar.

A first set of experiments focused on the influence of the variation of the oxygen concentration on conversion and selectivity (see Figure 7-19). The corresponding conversions increased only slightly on increasing the oxygen content from 10 to 50 %, whereas selec-

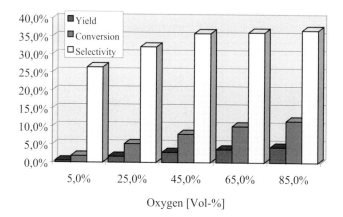

Fig. 7-19. Conversion, selectivity and yield of the partial oxidation of ethylene at varying oxygen concentrations. Ethylene concentration, 3 %; pressure, 5 bar; gas flow, 4 l/h; temperature, 277 °C (by courtesy of Schüth, MPI Mülheim).

tivity showed a more pronounced increase up to a maximum at 40 %. Selectivities of up to 50 % were found, conversions ranged from 10 to 12 % and the maximum yield was about 6 %. In addition, the experimental experiences confirmed safe handling of processes with parameters potentially being in the explosive range, e.g. an ethylene concentration of 3 %, a pressure of 5 bar, a gas flow of 4 l/h and a temperature of 277 °C.

By lowering the reaction temperature, it was possible to increase further the selectivities of the microreactor up to 65 %. This value is higher than that for polycrystalline silver catalysts, but still lower than the industrial benchmark of up to 80 %. As a measure of the accuracy of the data obtained, a determination of the activation energy amounted to 48 kJ/mol, which is comparable to data in literature.

A further optimization of process parameters led to duplication of the conversion to over 24 % at nearly constant selectivity, but at the expense of a decrease in the reaction rate. Applying higher gas flow rates and ethylene concentrations, resulted in the opposite behavior, namely a decrease of conversion, but enhancement of reaction rate. Increasing the pressure to higher than 5 bar was not a suitable strategy for performance optimization of the microreactor, because conversions and selectivities turned out to be constant. The optimum reaction temperature was found to be 267 °C.

Catalyst activity was maintained over an operating time of more than 1000 h. However, roughening and pitting phenomena of the surface could be observed (see Figure 7-20) as already described for other silver catalysts [53].

Concluding Remarks

The performance of the ethylene oxide microreactor, although outlined by first benchmarking experiments only, reached comparative values to polycrystalline silver catalysts and is not far from the corresponding industrial benchmark. Thus, the technical potential of microreactors for on-site and on-demand production of ethylene oxide in the l/h range was demonstrated. Commercialization of microreactors for such a field of application will depend on economic factors, a proper analysis being needed in near future.

Closer to commercialization certainly is the use of the microreactor as a measuring tool, in particular for processing in regimes otherwise not accessible, e.g. explosive regimes.

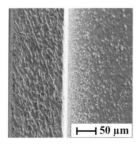 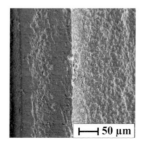

Fig. 7-20. Surfaces of the silver catalyst platelets of the microreactor before (left) and after (right) 1000 h of processing using the partial oxidation of ethylene [1].

Regarding this application, the results discussed above clearly demonstrated a safe handling of volatile toxic substances as well as concrete advantages with respect to safety.

7.6 Microsystems with Integrated Catalyst Structures, Heat Exchanger and Sensors

7.6.1 Oxidation of Ammonia

A research team at the Massachusetts Institute of Technology (MIT), Cambridge (USA) developed a microreactor design allowing detailed kinetic studies of process regimes, being extended with respect to macroscopic reactors, and fast process control of gas phase reactions [4, 13, 54–56]. One focus of the work was the application of simulation tools for the prediction of temperature profiles within the microreactor in order to optimize its design [55]. Theoretical results were accompanied by experimental data gained from miniaturized temperature and flow sensors placed close to the reaction volume [4, 13, 54].

In a joint research program of MIT with DuPont, Wilmington (USA), benefits due to design optimization and advanced process control of the microreactor were demonstrated for several examples of use, all belonging to the class of partial oxidations. Typical of these reactions is that the desired products belong to intermediate species, while side and consecutive reactions, resulting in total oxidation, often yield thermodynamically more stable products. A characteristic feature of partial oxidation processes is an ignition/extinction behavior, i.e. the reactions typically start at a certain temperature level only, referred to as ignition, and from then on rise steeply in temperature. When the reaction temperature is decreased, the opposite behavior occurs, referred to as extinction. Most results of the MIT/DuPont work were published for the oxidation of ammonia to various nitrogen oxides, supplemented by methane partial oxidation studies [56].

$$NH_3 \quad \xrightarrow[\text{Pt}]{O_2} \quad NO, N_2O$$

T-Shaped Microreactor equipped with Sensors and Heaters
A microreactor was equipped with two flow sensors in the entrance region as well as five heating units and ten temperature sensors on top of the long axis of a T-shaped channel structure, i.e. the reaction volume (see Figure 7-21). The shape of all sensors and heaters, being deposited by means of thin film technologies, was similar to meandering stripes. A thin catalyst layer was coated, using the same techniques, on the opposite side of the channel wall carrying the sensors and heaters. Temperature control was performed before and behind the extension of each heating element, by the so-called up- and downstream temperature sensors.
A unique feature of the ammonia oxidation reactor is the close control over temperature distribution achievable by means of heat conduction steered by a proper choice of materials

209

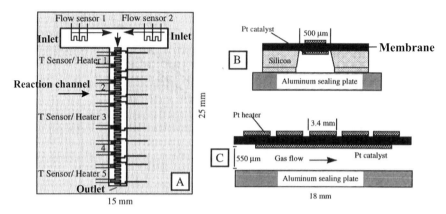

Fig. 7-21. Schematic representation of a T-shaped microreactor manufactured by means of silicon micromachining and thin film technology. A) top view; B) cross-section of reaction channel, perpendicular to flow direction; C) cross-section of reaction channel, parallel to flow direction [4].

and reactor geometry. In the course of the studies it turned out that heat conduction via the reactor material is the most prominent heat transfer mechanism, whereas heat convection, e.g. by the reactant flow or losses to the environment, is of minor importance. Concerning the heat transfer properties of the reaction channel, two entities of wall construction materials could be identified, namely a bulk silicon part based on an aluminum plate, acting as a heat sink, and a filigree membrane, covering the silicon channel walls, of variably tunable heat removal properties.

Modifications of the Catalyst Carrying Membrane
In order to control heat removal, this membrane was realized in two modifications, a thin version composed of silicon nitride and a thicker one made of silicon. The distinct difference in heat conductivity of the micro-structured materials (silicon 140 W/m K, silicon nitride 14 W/m K) and in membrane thickness (silicon 2.6 µm, silicon nitride 1 µm) results in notably different heat conduction properties. In case of an additional source of thermal energy, e.g. from heat release of an exothermic reaction, the low conduction of the silicon nitride membrane leads to hot-zone formation. This temperature deviation may occur in both the axial and radial directions. The silicon membrane, instead, has far better heat transfer characteristics, allowing removal of the excess heat.

In the T-shaped channel structure of size 15 mm x 25 mm, the reactants are mixed by entering a counter-flow configuration and are guided therefrom into a longer straight reaction channel (see Figure 7-20). This channel has a width and a depth of about 500 µm. The bottom of the channel structure, sealed with the aluminum plate, was equipped with openings for reactant feed and product outlet. The reaction mixture was analyzed by means of a mass spectrometer.

Microfabrication

Fabrication of the microreactor was exemplarily reported in the case of the silicon membrane microreactor. A silicon-on-insulator (SOI) wafer was coated with low-stress silicon nitride. After patterning and plasma etching of the nitride, the channel structure was etched with KOH to a depth smaller than the final value to allow easy IR alignment for the following coating step of platinum as heater and sensor material. The latter was achieved by lift-off photoresist patterning and subsequent e-beam deposition of platinum. Thereafter, the silicon was etched to the final depth, this process being stopped at a buried oxide layer. This material was removed by a buffered oxide etch (BOE) process exposing the stress-free and non-deformed silicon membrane. Thereafter, the platinum catalyst was deposited selectively in the microchannels by means of an e-beam evaporation process using a shadow-mask.

Simulations of Thermal Behavior

Simulations were performed to assist the microreactor design by providing an efficient base for thermal management. It was aimed to construct a system with a heat removal rate larger than the rate of heat generation, e.g. by reaction. However, to avoid superfluous heat consumption of the microreactor by setting the excess of heat transfer rates too high, precise design criteria based on simulations had to be developed to allow fine-tuning of this figure.

Simulations were based on a finite element method (FEM) and included a simultaneous solution of momentum, energy and mass conservation. Geometric boundary conditions were taken fully into account incorporating three-dimensional simulation and including the transport phenomena in the gaseous mixture as well as the kinetic data of the reaction. A comparison of the simulation results derived from temperature sensor measurements with a non-reacting gas allowed it to be judged if this methodology is really suited as a precise tool for microreactor design. Monitoring temperature profiles as a function of the heater power and the gas flux demonstrated excellent agreement of theroretical and experimental data [55].

Measured Temperature Profiles

In the framework of extensive investigations, temperature values were determined as a function of channel width, channel length, time, heater input power, etc. Analyzing a radial temperature profile within the membrane, it was found that a hot-zone was generated, i.e. while in the middle about 250 °C was measured, the edges of the membrane remained close to room temperature (see Figure 7-22). This behavior is due to the surrounding silicon channel walls acting as a heat sink, thereby efficiently removing the heat at the membrane edges.

Thus, the design of the heater had to be adjusted in order to achieve a much more uniform radial temperature distribution. The existing meandering heater design, covering nearly the whole width of the membrane, was replaced by a two-heater design, based on heating stripes at the membrane edges only. The resulting uniform temperature profile was characterized by a nearly flat plateau which steeply decreased to the membrane edges. It could be

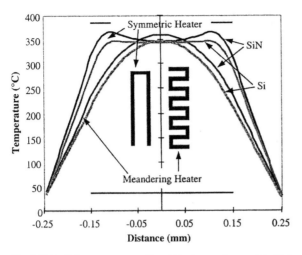

Fig. 7-22 Radial temperature profile within the membrane of the T-shaped gas phase microreactor, as determined by thin film deposited thermal sensors [4].

further shown that this type of profile did not change significantly on increasing the gas flow, which slightly decreased the temperature in the middle of the membrane by heat convection.

Further experiments were undertaken to investigate the axial and radial temperature distribution within the microchannel [4]. It could be shown that the highest temperature was obtained directly at the membrane, i.e. at the catalyst site, while a steep drop in temperature occurred along the depth of the microchannel. Again, this clearly illustrates the action of the surrounding channel walls as heat sinks. Thus, a virtually cold reaction gas is exposed at its boundary layer to a hot, reactive catalyst zone. A case study focused on the efficiency of mass transfer to this reaction site [55].

Oxidation of Ammonia
On the basis of these comprehensive theoretical and experimental investigations, the oxidation of ammonia was utilized as an example of use to demonstrate the benefits of improved thermal control by the microreactor. The experiments were focused on the feedback of reaction temperature on the input of heat power, i.e. the ignition–extinction behavior. Comparing this figure for two microreactors equipped with the silicon and silicon nitride membranes, a very interesting result was obtained, highlighting the different heat dissipation properties.

In the silicon nitride case, the temperature versus power curve was characterized by a hysteresis loop (see Figure 7-23) [13, 54]. With increasing input power the temperature first rises linearly and, at a distinct input power, jumps to a much higher value due to ignition. On reducing the input power, temperature decreases but at a higher level as compared with ignition until it jumps back to the previous curve. Due to the higher heat re-

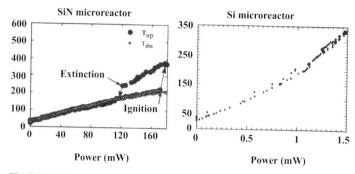

Fig. 7-23. Influence on the reaction characteristics for ammonia oxidation in two different microreactors with low (left) and high (right) heat dissipation near the catalyst zone demonstrated by means of input power versus temperature curves [54].

moval, a sufficiently thick silicon membrane does not show such a hysteresis loop but rather a continuous increase of temperature with input power. A decrease in silicon membrane thickness results in an intermediate type of behavior mentioned above [13], yielding a hysteresis loop, but with significantly changed ignition, extinction and upper steady-state temperatures.

For corresponding boundary conditions, the different behavior of the two reactor types results in a compositional change of the reaction products of the Pt-catalyzed ammonia oxidation. NH_3 and O_2 mainly react to give N_2 and NO in the silicon nitride membrane reactor while, in addition to N_2 and NO, significant amounts of N_2O were observed in the silicon membrane reactor. The relative amounts of NO and N_2 generated differ with respect to the type of microreactor as well. A plot of NO/N_2 selectivity versus temperature reveals a steep rise above 550 °C in case with the silicon nitride membrane [54]. Using silicon membranes, only a small increase of this figure in the temperature range from 350 to 550 °C is observed.

Use of T-Shaped Microreactor as Laboratory Measuring Tool
Apart from affecting the product composition, i.e. raising production issues, the above-mentioned characteristics render the microreactor a useful laboratory tool for the studying kinetics of rate-limiting processes in exothermal partial oxidation reactions. Conventional reactors, in some cases, fail to give precise data, e.g. on reactant conversion and product selectivity, due to mass transfer limitations caused by the enormous temperature rise after ignition. In the case of using the microreactor, such limitations virtually do not exist for a broad range of operating conditions. This was impressively demonstrated, in the course of the ammonia oxidation experiments, by monitoring ammonia conversion versus temperature. Data could be gained at intermediate temperatures, otherwise not accessible, and showed high accuracy, e.g. no hysteresis, when comparing results of increasing and decreasing temperature cycles.

A further aspect of the studies was focused on the improved process control achievable in microreactors. The thermal response time was revealed in open-loop experiments and compared with the time-scale of typical residence times. It was found that time constants less than 20 ms are needed to reach the reaction temperature. By implementing a temperature control scheme, further improvements were realized, e.g. feedback control with thermal response times of about 1 ms was achievable. Since this value is smaller than typical residence times of gas flows in microreactors, ranging from a few tens of milliseconds to seconds, this experiment proved the potential of using microreactors to operate in a dynamic fashion.

Concluding Remarks
All the investigations clearly showed that microreactors allow either reaction conditions to be achieved which are not feasible in macroscopic devices or at least enable the precision of setting of operating parameters to be considerably improved. The silicon reactor, for example, can be operated in a regime between ignition and extinction temperatures which results in reaction products different from those of a macroscopic device. This possibility of changing the reaction products may be of major importance also for a number of other technically interesting reactions such as ethylene oxidation or generally for reactions in the explosive regime.

Suprisingly good agreement between measurements and simulations demonstrates that, at least for relatively simple devices and reactions, a theoretical prediction of the behavior of microreactors and, accordingly, a detailed optimization are possible. This profitable and reliable use of simulation results, in combination with the precise setting of operating parameters and excellent process control by fast response times, renders such microreactors a versatile laboratory tool for kinetic investigations.

7.6.2 H$_2$/O$_2$ Reaction

Use of Membrane Technology for Reactant Separation in Microreactors
Membrane reactors provide a new concept for chemical processing, characterized by the interplay between gas separation of reactant streams and chemical reaction. By continuous separation of reactants and products, non-equilibrium chemistry can be established. A further consequence of combining separation and reaction is higher flexibility concerning the use of feed streams and, in some cases, selectivity of the whole process may be improved. A particularly interesting field is the performance of hydrogenations/dehydrogenations in palladium membrane reactors. Current systems mainly lack thick palladium films, rendering them too expensive and decreasing the efficiency.

In order to overcome these limitations of this innovative concept, researchers at the Massachusetts Institute of Technology (MIT), Cambridge (USA) developed a microreactor equipped with thin palladium coatings [57]. The microreactor design was oriented on the T-shaped silicon channel structure, equipped with miniaturized heater and sensors, originally developed for the oxidation of ammonia [4, 13]. However, since two flow channels

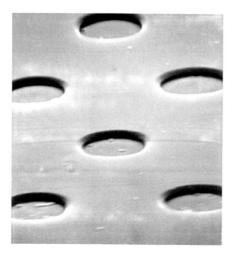

Fig. 7-24. Scanning electron micrograph of a free-standing palladium membrane within the holes of a thin sheet [57].

were separated by a membrane are needed, a structurally similar channel layer made in polymer was set on top of the silicon channel layer. The two channels are separated by a thin sheet composed of silicon oxide and nitride layers. Below this sheet a thin palladium membrane was formed by coating, forming the top side of the underlying silicon channel. In order to allow gas flow from the polymeric to the silicon channel, the silicon oxide/ nitride sheet contains an array of many microstructured holes. Within the holes the thin palladium film is free standing (see Figure 7-24).

Microfabrication
Microfabrication was achieved by standard silicon micromachining based on lithography, thin film deposition and wet and dry etching techniques. The microchannels are 12 mm long and 700 μm wide. The whole device amounts to 16 mm x 8 mm. For generation of the microholes, a dry nitride etch of the silicon nitride layer was utilized. The palladium layer was deposited on a thin titanium adhesion layer. Finally, the oxide layer was patterned using wet etching, thereby releasing the palladium membrane. The polymer channel structure was realized by molding. A negatively shaped copy was achieved from a silicon master. A capillary was introduced into the copy structure which thereafter was cast with an epoxy polymer.

Membrane Permeability and Proof of Feasibility for H_2/O_2 Reaction
Before carrying out a test reaction, the mechanical strength and hydrogen flux achievable were determined. The free-standing thin palladium membrane could withstand relatively high pressure gradients of up to 5 atm. A selective hydrogen flux through the membrane

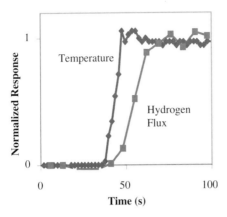

Fig. 7-25. Fast response of a palladium membrane reactor due to its low mass and adjusted thermal properties, as evidenced by increase of hydrogen flow through the membrane [57].

with a separation factor of 1800 was realized. In accordance with theory, the hydrogen flux increased with rising temperature and pressure gradients. At an operating temperature of 500 °C, a flux of 600 sccm/cm^2 was measured. Low mass and good thermal insulation, due to the oxide/nitride layers, of the membrane result in short response times down to 10 ms (see Figure 7-25).

$$2\,H_2 + O_2 \xrightarrow[Pd]{} 2\,H_2O$$

The performance of the device was evidenced by carrying out the H_2/O_2 reaction. A nitrogen/hydrogen mixture was passed through the silicon channel, while air was used as fluid on the other side of the palladium membrane. After penetrating the heated membrane, a reaction of hydrogen with oxygen occurred. This was evidenced by water condensing in the colder parts of the microreactor.

Concluding Remarks
The work of the MIT team demonstrated elegantly that further benefits can be expected when adapting innovative concepts developed for macroscopic reactors to microstructured devices. The small dimensions of microdevices facilitate transport phenomena to and across the membrane, thereby allowing high specific fluxes and increasing functionality due to micro-scale integration of several operations. Such highly integrated microreactors may be one of the favorite designs of next generation devices. The construction of standardized modules carrying out a single operation only is an alternative thereof.

7.7 Microsystems with Integrated Mixer, Heat Exchanger, Catalyst Structures and Sensors

7.7.1 HCN Synthesis via the Andrussov Process

Fast High-Temperature Process Generating Hazardous Gases
Researchers at the Institut für Mikrotechnik Mainz (Germany) developed a high-temperature microreactor for millisecond synthesis of small gas volumes in the framework of a contract research project [51, 58, 59], founded by the chemical companies Axiva (formerly Aventis Research and Technologies), DuPont, Degussa, BASF, Hüls and Rhodia/Rhône-Poulenc. Their common interest was the realization of a compact device capable of producing hazardous gases by means of a fast process.

The investigations were dedicated to basic R&D work and the process chosen, the Andrussov process, was used as a test reaction to prove feasibility only. However, in case of success, it was aimed to transfer the results into commercial application. Potential examples of use for this microreactor are the determination of reaction kinetics and production on-site and on-demand.

The studies should further reveal the benefits of a uniform heat distribution, i.e. isothermal process conditions, and short contact times for a well-known high-temperature reaction. Thus, the project was of precompetitive nature, combining both basic research issues with industrial implementation.

The formation of hydrogen cyanide from oxidative dehydrogenation of methane and ammonia, i.e. by means of the Andrussov process, was selected for this purpose [60]. This reaction fulfills nearly all criteria required for a process to benefit from miniaturization, namely a high reaction rate, exothermicity and mechanistic complexity. The latter is due to multiple reaction pathways including fast side and consecutive elemental reactions. In addition, detailed know-how concerning the Andrussov process has been collected in a number of scientific and application oriented studies over more than seven decades [61–65], thus being a suitable benchmark.

$$CH_4 + NH_3 + 1.5\,O_2 \xrightarrow[\text{Pt}]{} HCN + 3\,H_2O$$

Construction of Microreaction System
The construction of the total microreaction system was complex with respect to the number of components assembled and their different materials. The system consists of a stainless steel preheater/mixer, platinum catalyst structure on a ceramic support, a stainless steel heat exchanger, and an in-line temperature sensor being part of the platinum catalyst structure (see Figure 7-26). Design criteria of the microreactor were adapted so that operating conditions corresponding to those of the traditional Andrussov process could be applied.

In a block equipped with heating cartridges methane, ammonia, and oxygen, flowing in three separate tubes, are preheated up to 600 °C. The diameter of these tubes is decreased in

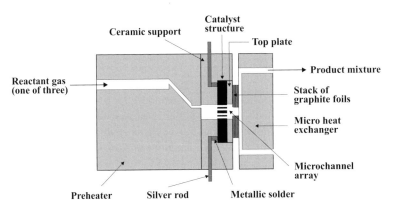

Fig. 7-26. Schematic of the microreactor assembly. The design is only schematically presented, in particular, concerning the dimensions of the components [59].

several steps, resulting in holes of 60 μm diameter at the end of the block. Thereafter, the gases enter a diffusion chamber and, according to calculations, are completely mixed after 3 mm. The reactant mixture passes 24 boreholes of 70 μm width in a 250 μm thick platinum catalyst structure inserted in a ceramic support. After a short entrance length, the temperature of the gas is expected to adapt to more than 1000 °C hot wall of the boreholes, as suggested by calculations for this geometry and for similar applications.

In addition to these heat transfer considerations, it was checked if mass transfer from the gas volume to the catalytic surface is sufficient. Calculations confirm that a single gas molecule will sustain at least one collision with the wall of the boreholes, potentially leading to a catalyzed conversion to the product molecules. By varying the volume flow, residence times in the catalyst structure between 0.1 and 1 ms were achievable. Thereafter, the hot product gas mixture passes an interconnecting zone of 300 μm length, and is cooled to 120 °C in a micro heat exchanger. Using air as the coolant medium, calculations predict cooling times of about 100 μs. It could be experimentally verified that the gas is cooled to 120 °C after passing this heat exchanger.

The disassembled microreactor, including the above-mentioned components as well as heating cartridges, insulation jackets, graphite flat sealings and supplementary fluidic equipment, is shown in Figure 7-27.

Catalyst Structure and Micro Heat Exchanger
Despite the close resemblence of operating parameters achievable in the microreactor compared with the large-scale Andrussov reactor, two differences have to be mentioned. First, the distance between the hot reaction zone and the cold heat exchanger was minimized by integration of both components within the microreactor. For the waste-heat boiler used in the large-scale reactor no cooling rate or distance from the catalytic meshes was specified in the literature [64, 65]. Nevertheless, it can be assumed that the corresponding data of the micro heat exchanger are improved.

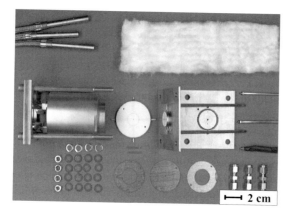

Fig. 7-27. Photograph of the microstructured components, heating cartridges, insulation jackets, graphite flat sealings, and supplementary fluidic equipment of the high-temperature microreactor [59].

The second difference refers to the catalytic zone. While the industrial process is carried out on several layers of catalytic meshes, the microreactor utilizes a multichannel configuration within the catalyst structure. At the meshes, a self-sustaining reaction is initiated, and the microstructured catalyst, instead, has to be externally heated by passing electrical current through thin walls separating the microchannels.

These channels are split into two groups, each consisting of two rows with five channels of about 70 μm diameter (see Figure 7-28). In this highly symmetrical arrangement, each channel is surrounded by three thin channel walls which act as heat sources when passing electric current generated by an electrical power supply. According to heat flux calculations, a uniform temperature profile is achieved, with a maximum temperature deviation of less than 1.5 K. In order to reach temperatures of more than 1000 °C, the thickness of the channel walls had to be restricted to 40 μm.

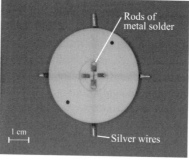

Fig. 7-28. Reaction zone consisting of a platinum multichannel catalyst structure (left) and a ceramic support made of aluminum oxide (right) [59].

219

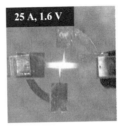

Fig. 7-29. Glowing catalyst structure in air [51].

Figure 7-29 demonstrates that the operational parameters for resistance heating were sufficiently set to realize the reaction temperature. When contacted with a power supply, only the multichannel configuration in the middle of the catalyst structure glows, hence having a much higher temperature than the residual parts of the platinum structure. In addition, the efficiency of thermal insulation of the ceramic support was evidenced by a steep temperature drop of about 700 to 800 °C between the border of the support and the heated microchannels.

Measurements with thin thermocouples attached to the channel configuration were taken as a first measure of the actual temperature in the channels. Thereby, the attainment of temperatures higher than 1000 °C was confirmed. Later, temperatures could be determined by following the temperature induced change of electrical resistivity using an in-line temperature sensor. A calibration curve allowed the determination of the reaction temperature even during processing inside the microreactor.

After passing the reaction zone, the product gas was guided to the micro heat exchanger. From a centrally located entrance area the product gas was fed to four microchannels which form a cross-type structure (see Figure 7-30). In order to cool the gas efficiently, channels with a width of 60 μm and a depth of 500 μm were chosen to provide large contact surfaces to the cooling channels carrying the heat exchange medium air. Each product channel is surrounded by two cooling channels operating in a counter-flow mode. After passing the four product channels, the product gas is recollected and guided through a borehole to an outlet tube.

Microfabrication
Fabrication of the contours of the preheater and heat exchanger was performed by conventional precision engineering. The outlet holes of the mixing structure were drilled by means

220

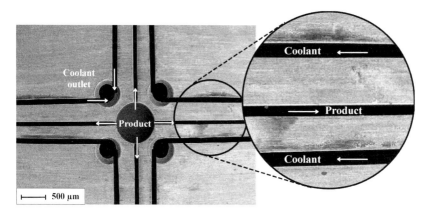

Fig. 7-30. Cross-type micro heat exchanger with large contact surfaces for operation in co- or countercurrent flow mode [59].

of micro thin-wire erosion, the cooling microchannels were realized by a die sinking process. Special efforts were made for the microfabrication of the catalyst structure. The contours were cut out of a thin platinum foil using micro thin-wire erosion. By drilling with tiny wire-like electrodes of about 40 μm, the microchannels were realized in the center of the catalyst structure.

The ceramic support was manufactured by a combination of several ultraprecision machining techniques [51]. In a first approach, the catalyst structure was irreversibly embedded in a ceramic glue using casting techniques. A second approach achieved a fluid tight sealing by reversibly covering the catalyst structure with a ceramic top plate and a graphite sealing, both being laser micromachined. Within the ceramic support four holes were drilled and filled with silver rods for electrical connection to a power supply. These rods were joined to a commercial metal solder which also allowed a gas-tight interconnection of the metal/ceramic interfaces. The solder part was structured by micromilling to yield pads for reversible insertion of the catalyst structure (see Figure 7-31).

Experimental Tests Proving Feasibility
Two identically assembled microreactors were tested in parallel by two industrial partners, BASF and Rhodia/Rhône-Poulenc. A complete experimental set-up was realized at IMM integrating the microreactor, several fluid connecting lines and boxes for electrical contact on a support plate with a size half of that of a standard laboratory table. Analysis of the product spectrum was performed using gas chromatography.

Figure 7-32 shows a typical product composition of selected gases, given in vol.-%, obtained at BASF, including the product HCN, side and follow-up products (CO, CO_2) and two of the reactants (NH_3, CH_4). The volumetric content of the product is proportional to the yield and the amount of remaining reactant is inversely proportional to the conversion.

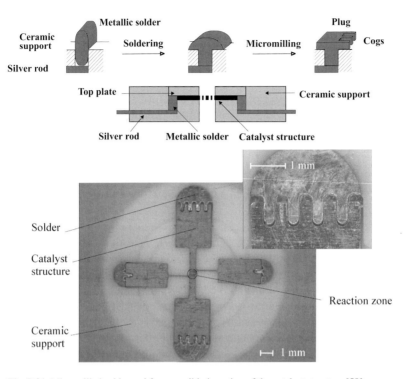

Fig. 7-31. Micromilled solder pad for reversible insertion of the catalyst structure [59].

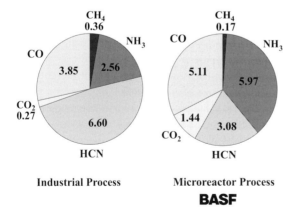

Industrial Process   **Microreactor Process**

BASF

Fig. 7-32. Comparison of reactor performance of the microreactor, based on measurements of BASF, with results of the industrial Andrussov process [59].

222

Other gases of the product mixture, namely O_2, N_2 and H_2, were omitted for reasons of clarity. As a technical benchmark, a standard composition of a product gas obtained in the industrial Andrussov process is included [63–65]. The results of Rhône-Poulenc are discussed elsewhere [59].

A significant formation of hydrogen cyanide with yields of up to 31 % was observed, nearly half the value of the industrial process [56–58]. The two processes, however, differ significantly with respect to methane selectivity. In the case of the microreactor a notable formation of carbon side and follow-up products was observed, referring to carbon monoxide and carbon dioxide. The methane conversion is close to 100 % for both the large-scale and microreactor. The ammonia selectivity of about 60 % is close to the industrial value, also underlining the principal feasibility of the microreactor concept.

Comparison with Monolith Reactors
Another benchmark for reactor efficiency is revealed by a comparison with results of monolithic reactors which are structurally analogous to the microreactors, but contain larger channels of about 500 to 1000 μm diameter [61]. When operated in the laminar regime, the former class of reactors gave only low product yields and low conversions of methane and ammonia, obviously due to insufficient mass transfer (see Figure 7-33). Only a ceramic monolith operating in the turbulent regime gave somewhat better results, exceeding the yield measured in the microreactor by 7 %.

The superior reactor performance of the microreactor compared with the laminar monoliths is mainly caused by a 7–14 times smaller hydraulic channel diameter and, hence, shorter diffusional distances. However, the microreactor performance is worse than the

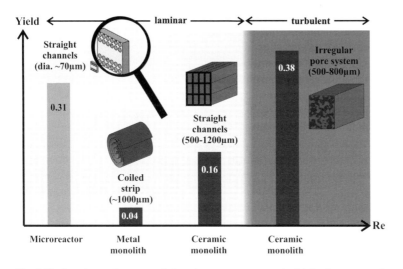

Fig. 7-33. Superior performance of the microreactor compared with laminar operated metallic and ceramic monoliths. Data for a turbulent operated ceramic monolith are included [59].

turbulent monolith. One possible explanation for this behavior relies on benefits from a longer ratio of channel length to width of the monolith [66], e.g. ranging from about 10 to 30, whereas the corresponding figure for the microreactor is only about 3.5.

Concluding Remarks

Due to stringent time limits of the industrial investigations, the potential of the microreactor could only partially be evaluated. Experiments were performed only within a limited parameter range, in particular concerning residence time and temperature variation. Nevertheless, the work laid the foundation for an adaptation of microfabrication techniques in order to realize microfluidic components made of different materials, including ceramics, which can withstand high temperatures. This especially holds for microstructuring by ultraprecision and micro electro discharge machining.

The microreactor concept presented above was able to extend significantly the temperature range of existing gas phase microreactors. Following studies on another high-temperature microreactor carrying out millisecond reactions [14], the feasibility of sub-millisecond operation was demonstrated within the studies discussed here. Although both concepts need further design improvement and other conceptual modifications, they should be regarded as important first steps of an on-going development towards mobile, small-scale reaction units of flexible capacity.

7.8 References

[1] Kestenbaum, H., Lange de Oliveira, A., Schmidt, W., Schüth, F., Gebauer, K., Löwe, H., Richter, T.; *"Synthesis of ethylene oxide in a catalytic microreacton system"*, in Ehrfeld, W. (Ed.) *Microreaction Technology: 3rd International Conference on Microreaction Technology, Proceedings of IMRET 3*, pp. 207–212, Springer-Verlag, Berlin, (2000).

[2] Wießmeier, G., Hönicke, D.; *"Strategy for the development of micro channel reactors for heterogenously catalyzed reactions"*, in Ehrfeld, W., Rinard, I. H., Wegeng, R. S. (Eds.) *Process Miniaturization: 2nd International Conference on Microreaction Technology; Topical Conference Preprints*, pp. 24–32, AIChE, New Orleans, USA, (1998).

[3] Wießmeier, G., Hönicke, D.; *"Microfabricated components for heterogeneously catalysed reactions"*, J. Micromech. Microeng. **6**, pp. 285–289 (1996).

[4] Jensen, K. F., Hsing, I.-M., Srinivasan, R., Schmidt, M. A., Harold, M. P., Lerou, J. J., Ryley, J. F.; *"Reaction engineering for microreactor systems"*, in Ehrfeld, W. (Ed.) *Microreaction Technology, Proceedings of the 1st International Conference on Microreaction Technology; IMRET 1*, pp. 2–9, Springer-Verlag, Berlin, (1997).

[5] Liauw, M., Baerns, M., Broucek, R., Buyevskaya, O. V., Commenge, J.-M., Corriou, J.-P., Falk, L., Gebauer, K., Hefter, H. J., Langer, O.-U., Löwe, H., Matlosz, M., Renken, A., Rouge, A., Schenk, R., Steinfeld, N., Walter, S.; *"Periodic operation in microchannel reactors"*, in Ehrfeld, W. (Ed.) *Microreaction Technology: 3rd International Conference on Microreaction Technology, Proceedings of IMRET 3*, pp. 224–234, Springer-Verlag, Berlin, (2000).

[6] Wießmeier, G., Schubert, K., Hönicke, D.; *"Monolithic microstructure reactors possessing regular mesopore systems for the successful performance of heterogeneously catalyzed reactions"*, in Ehrfeld, W. (Ed.) *Microreaction Technology, Proceedings of the 1st International Conference on Microreaction Technology; IMRET 1*, pp. 20–26, Springer-Verlag, Berlin, (1997).

[7]   Richter, T., Ehrfeld, W., Erntner, D., Gebauer, K., Golbig, K., Günther, N., Hang, T., Hausner, O., Löwe, H., Schiewe, J., Scholl, T., Zapf, R.; *"Entwicklung und Realisierung von Mikroreaktoren und ihren Komponenten für Gasphasenreaktionen"*, in Proceedings of the "MicroEngineering 99", 29 Sept. – 1 Oct., 1999; pp. 288–297; Stuttgart.

[8]   Results of IMM, unpublished

[9]   Fichtner, M., Benzinger, W., Hass-Santo, K., Wunsch, R., Schubert, K.; *"Functional coatings for microstructure reactors and heat exchangers"*, in Ehrfeld, W. (Ed.) *Microreaction Technology: 3rd International Conference on Microreaction Technology, Proceedings of IMRET 3*, pp. 90–101, Springer-Verlag, Berlin, (2000).

[10] Pfeifer, P., Fichtner, M., Schubert, K., Liauw, M. A., Emig, G.; *"Microstructured catalysts for methanol-steam reforming"*, in Ehrfeld, W. (Ed.) *Microreaction Technology: 3rd International Conference on Microreaction Technology, Proceedings of IMRET 3*, pp. 372–382, Springer-Verlag, Berlin, (2000).

[11] Sainz, M. A., Torrecillas, R., Moya, J. S.; J. Am. Ceram. Soc. **76**, 7 (1993) 1869.

[12] Bhave, R. R.; *Inorganic Membranes,* Van Norstrand Reinhold, New York (1991).

[13] Franz, A. J., Ajmera, S. K., Firebaugh, S. L., Jensen, K. F., Schmidt, M. A.; *"Expansion of microreactor capabilities trough improved thermal management and catalyst deposition"*, in Ehrfeld, W. (Ed.) *Microreaction Technology: 3rd International Conference on Microreaction Technology, Proceedings of IMRET 3*, pp. 197–206, Springer-Verlag, Berlin, (2000).

[14] Tonkovich, A. L. Y., Zilka, J. L., Powell, M. R., Call, C. J.; *"The catalytic partial oxidation of methane in a microchannel chemical reactor"*, in Ehrfeld, W., Rinard, I. H., Wegeng, R. S. (Eds.) *Process Miniaturization: 2nd International Conference on Microreaction Technology; Topical Conference Preprints,* pp. 45–53, AIChE, New Orleans, USA, (1998).

[15] Fichtner, M., Benzinger, W., Schubert, K.; *"Fouling in microchannel devices"*, "4th International Conference on Microreaction Technology, IMRET 4", 5–9 March, 2000; Atlanta, USA.

[16] Cybulski, A., Moulijn, J. A.; *"Monoliths in heterogeneous catalysis"*, Catal. Rev. – Sci. Eng. **36**, 2, pp. 179–270 (1994).

[17] Schubert, K., Bier, W., Brandner, J., Fichtner, M., Franz, C., Linder, G.; *"Realization and testing of microstructure reactors, micro heat exchangers and micromixers for industrial applications in chemical engineering"*, in Ehrfeld, W., Rinard, I. H., Wegeng, R. S. (Eds.) *Process Miniaturization: 2nd International Conference on Microreaction Technology, IMRET 2; Topical Conference Preprints,* pp. 88–95, AIChE, New Orleans, USA, (1998).

[18] Wießmeier, G., Hönicke, D.; *"Microreaction technology: development of a micro channel reactor and its application in heterogenously catalyzed hydrogenations"*, in Ehrfeld, W., Rinard, I. H., Wegeng, R. S. (Eds.) *Process Miniaturization: 2nd International Conference on Microreaction Technology, IMRET 2; Topical Conference Preprints,* pp. 152–153, AIChE, New Orleans, USA, (1998).

[19] Wießmeier, G.; *"Monolithische Mikrostruktur-Reaktoren mit Mikroströmungskanälen und regelmäßigen Mesoporensystem für selektive, heterogen katalysierte Gasphasenreaktion"*, Dissertation; Karlsruhe, (1996).

[20] Hönicke, D., Wießmeier, G.; *"Heterogeneously catalyzed reactions in a microreactor"*, in Ehrfeld, W. (Ed.) *Microsystem Technology for Chemical and Biological Microreactors, DECHEMA Monograph,* Vol. 132, pp. 93–107, Verlag Chemie, Weinheim, (1996).

[21] Wießmeier, G.; *"Untersuchungen zur heterogen katalysierten Oxidation von Propen in einem Mikrostrukturreaktor"*, Diplomarbeit; Karlsruhe, (1992).

[22] Bier, W., Guber, A., Linder, G., Schaller, T., Schubert, K.; *"Mechanische Mikrotechnik – Verfahren und Anwendungen"*, KfK Ber. 5238, pp. 132–137 (1993).

[23] Wörz, O., Jäckel, K. P., Richter, T., Wolf, A.; *"Microreactors, a new efficient tool for optimum reactor design"*, in Ehrfeld, W., Rinard, I. H., Wegeng, R. S. (Eds.) *Process Miniaturization: 2nd International Conference on Microreaction Technology, IMRET 2; Topical Conference Preprints,* pp. 183–185, AIChE, New Orleans, USA, (1998).

[24] Hagendorf, U., Janicke, M., Schüth, F., Schubert, K., Fichtner, M.; *"A Pt/Al₂O₃ coated microstructured reactor/heat exchanger for the controlled H₂/O₂-reaction in the explosion regime"*, in Ehrfeld, W., Rinard, I. H., Wegeng, R. S. (Eds.) *Process Miniaturization: 2nd International Conference on Microreaction Technology; Topical Conference Preprints,* pp. 81–87, AIChE, New Orleans, USA, (1998).

[25] Walter, S., Frischmann, G., Broucek, R., Bergfeld, M., Liauw, M.; *"Fluiddynamische Aspekte in Mikroreaktoren"*, Chem. Ing. Tech. **71**, pp. 447–455, 5 (1999).

[26] Keller, S., Hunter, M. S., Robinson, D. L.; *"Structural features of oxide coatings on aluminium"*, J. Electrochem. Soc. **100**, p. 411 (1953).

[27] Jankowski, A. E., Matthee, T.; *"Teil 1: Das Färben von anodisiertem Aluminium mit organischen Farbstoffen unter den Aspekten Einfärbeverhalten und Lichtechtheit"*, Galvanotechnik **86**, 5 (1995) 1421.

[28] Jessup, G. R.; *"Troubleshooting aluminium anodizing; Part III: anodizing, coloring and sealing"*, Prod. Finish. 11, p. 52 (1992).

[29] Wießmeier, G., Hönicke, D.; *"Heterogeneously catalyzed gas-phase hydrogenation of cis, trans, trans-1,5,9-cyclododecatriene on palladium catalysts having regular pore systems"*, Ind. Eng. Chem. Res. **35**, (1996) 4412–4416.

[30] Schubert, K.; *"Entwicklung von Mikrostrukturapparaten für Anwendungen in der chemischen und thermischen Verfahrenstechnik"*, KfK Ber. **6080**, (1998) 53–60.

[31] Veser, G., Friedrich, G., Freygang, M., Zengerle, R.; *"A modular microreactor design for high-temperature catalytic oxidation reactions"*, in Ehrfeld, W. (Ed.) *Microreaction Technology: 3rd International Conference on Microreaction Technology, Proceedings of IMRET 3*, pp. 674–686, Springer-Verlag, Berlin, (2000).

[32] Kursawe, A., Dietzsch, E., Kah, S., Hönicke, D., Fichtner, M., Schubert, K., Wießmeier, G.; *"Selective reactions in microchannel reactors"*, in Ehrfeld, W. (Ed.) *Microreaction Technology: 3rd International Conference on Microreaction Technology, Proceedings of IMRET 3*, pp. 213–223, Springer-Verlag, Berlin, (2000).

[33] Silveston, P., Hudgins, R. R., Renken, A.; *"Periodic operations of catalytic reactors – introduction and overview"*, Catal. Today **25**, (1995) 91–112.

[34] Stepanek, F., Kubicek, M., Marek, M., Adler, P. M.; *"Optimal design and operation of a separating microreactor"*, Chem. Eng. Sci. **54**, (1999) 1494–1498.

[35] Rouge, A., Spoetzl, B., Gebauer, K., Schenk, R., Renken, A.; *"Microchannel reactors for fast periodic operation: the catalytic dehydration of isopropanol"*, in Proceedings of the "ISCRE 16, International Symposium on Chemical Reaction Engineering", 2000; p. in press; Cracow.

[36] Walter, S., Liauw, M.; *"Fast concentration cycling in microchannel reactors"*, in Proceedings of the "4th International Conference on Microreaction Technology, IMRET 4", pp. 209–214; 5–9 March, 2000; Atlanta, USA.

[37] Contractor, R. R.; *"Improved vapor Phase catalytic oxidation of butane to maleic anhydride"*, US 4 668 802, (26.05.1987); .

[38] Ehrfeld, W., Hessel, V., Löwe, H.; *"Extending the knowledge base in microfabrication towards chemical engineering and fluid dynamic simulation"*, in Proceedings of the "4th International Conference on Microreaction Technology, IMRET 4", pp. 3–22; 5–9 March, 2000; Atlanta, USA.

[39] Commenge, J. M., Matlosz, M., Corriou, J. P., Falk, L.; *"Modellierung, Optimierung und Regelung"*, Mikroreaktorsysteme in der chemischen Technik: Schriftliche Projektpräsentation zur 1. Projektgruppensitzung DECHEMA, Frankfurt/M (1999).

[40] Koubek, J., Pasek, J., Ruzicka, V.; *"Exploitation of an non-stationary kinetic phenomenon for the elucidation of surface processes in a catalytic reaction"*, New Horizons in Catalysis, pp. 853–862, Elsevier – Kodansha, Amsterdam-Tokyo, (1980).

[41] Creaser, D., Andersson, B., Hudgins, R. R., Silverston, P. L.; *"Transient kinetic analysis of the oxidative dehydrogenation of propane"*, J. Catal. **182**, pp. 264–269 (1999).

[42] Jäckel, K. P.; *"Microtechnology: Application opportunities in the chemical industry"*, in Ehrfeld, W. (Ed.) *Microsystem Technology for Chemical and Biological Microreactors*, Vol. 132, pp. 29–50, Verlag Chemie, Weinheim, (1996).

[43] Lerou, J. J., Harold, M. P., Ryley, J., Ashmead, J., O'Brien, T. C., Johnson, M., Perrotto, J., Blaisdell, C. T., Rensi, T. A., Nyquist, J.; *"Microfabricated mini-chemical systems: technical feasibility"*, in Ehrfeld, W. (Ed.) *Microsystem Technology for Chemical and Biological Microreactors*, Vol. 132, pp. 51–69, Verlag Chemie, Weinheim, (1996).

[44] Ehrfeld, W., Hessel, V., Haverkamp, V.; *"Microreactors"*, Ullmann's Encyclopedia of Industrial Chemistry, Wiley-VCH, Weinheim, (1999).

[45] Bier, W., Keller, W., Linder, G., Seidel, D., Schubert, K.; *"Manufacturing and testing of compact micro heat exchangers with high volumetric heat transfer coefficients"*, ASME, DSC-Microstructures, Sensors, and Actuators **19**, pp. 189–197 (1990).

[46] Ashmead, J. W., Blaisdell, C. T., Johnson, M. H., Nyquist, J. K., Perrotto, J. A., Ryley, J. F.; *"Integrated Chemical Processing Apparatus and Processes for the Preparation Thereof"*, EP 0688 242 B1, (19.03.1993); E.I. Du Pont de Nemours Co.

[47] Löhder, W., Bergmann, L.; *"Verfahrenstechnische Mikroapparaturen und Verfahren zu ihrer Herstellung"*, DD 246 257 A1, (21.01.1986); Akademie der Wissenschaften der DDR.

226

[48] Ehrfeld, W., Hessel, V., Möbius, H., Richter, T., Russow, K.; *"Potentials and realization of micro reactors"*, in Ehrfeld, W. (Ed.) *Microsystem Technology for Chemical and Biological Microreactors*, Vol. 132, pp. 1–28, Verlag Chemie, Weinheim, (1996).

[49] Kestenbaum, H., Lange de Olivera, A., Schmidt, W., Schüth, H., Ehrfeld, W., Gebauer, K., Löwe, H., Richter, T.; *"Synthesis of ethylene oxide in a catalytic microreactor system"*, in Proceedings of the "12th Int. Conf. on Catalysis 2000", 2000; Granada, Spain.

[50] Richter, T., Ehrfeld, W., Gebauer, K., Golbig, K., Hessel, V., Löwe, H., Wolf, A.; *"Metallic microreactors: components and integrated systems"*, in Ehrfeld, W., Rinard, I. H., Wegeng, R. S. (Eds.) *Process Miniaturization: 2nd International Conference on Microreaction Technology, IMRET 2; Topical Conference Preprints*, pp. 146–151, AIChE, New Orleans, USA, (1998).

[51] Löwe, H., Ehrfeld, W., Gebauer, K., Golbig, K., Hausner, O., Haverkamp, V., Hessel, V., Richter, T.; *"Microreactor concepts for heterogeneous gas phase reactions"*, in Ehrfeld, W., Rinard, I. H., Wegeng, R. S. (Eds.) *Process Miniaturization: 2nd International Conference on Microreaction Technology, IMRET 2; Topical Conference PreprintsProcess Miniaturization: 2nd International Conference on Microreaction Technology; Topical Conference Preprints*, pp. 63–74, AIChE, New Orleans, USA, (1998).

[52] Tsybulya, S. V., Kryukova, G. N., Goncharova, S. N., Shmakov, A. N., Bal'zhinimaev, B. S.; J. Catal. **154**, p. 194 (1995).

[53] Rehren, C., Muhler, M., Bao, X., Schlögl, R., Ertl, G.; Z. Phys. Chem. **174**, (1991) 11.

[54] Franz, A. J., Quiram, D. J., Srinivasan, R., Hsing, I.-M., Firebaugh, S. L., Jensen, K. F., Schmidt, M. A.; *"New operating regimes and applications feasible with microreactors"*, in Ehrfeld, W., Rinard, I. H., Wegeng, R. S. (Eds.) *Process Miniaturization: 2nd International Conference on Microreaction Technology; Topical Conference Preprints*, pp. 33–38, AIChE, New Orleans, USA, (1998).

[55] Quiram, D. J., Hsing, I.-M., Franz, A. J., Srinivasan, R., Jensen, K. F., Schmidt, M. A.; *"Characterization of microchemical systems using simulations"*, in Ehrfeld, W., Rinard, I. H., Wegeng, R. S. (Eds.) *Process Miniaturization: 2nd International Conference on Microreaction Technology; Topical Conference Preprints*, pp. 205–211, AIChE, New Orleans, USA, (1998).

[56] Srinivasan, R., I-Ming Hsing, Berger, P., Jensen, E. K. F., Firebaugh, S. L., Schmidt, M. A., Harold, M. P., Lerou, J. J., Ryley, J. F.; *"Micromachined reactors for catalytic partial oxidation reactions"*, AIChE J. **43**, pp. 3059–3069, 11 (1997).

[57] Franz, A. J., Jensen, K. J., Schmidt, M. A.; *"Palladium membrane microreactors"*, in Ehrfeld, W. (Ed.) *Microreaction Technology: 3rd International Conference on Microreaction Technology, Proceedings of IMRET 3*, pp. 267–276, Springer-Verlag, Berlin, (2000).

[58] Hessel, V., Ehrfeld, W., Golbig, K., Wörz, O.; *"Mikroreaktionssysteme für die Hochtemperatursynthese"*, GIT Labor-Fachz. 10, p. 1100 (1999).

[59] Hessel, V., Ehrfeld, W., Golbig, K., Hofmann, C., Jungwirth, S., Löwe, H., Richter, T., Storz, M., Wolf, A., Wörz, O., Breysse, J.; *"High temperature HCN generation in an integrated Microreaction system"*, in Ehrfeld, W. (Ed.) *Microreaction Technology: 3rd International Conference on Microreaction Technology, Proceedings of IMRET 3*, pp. 151–164, Springer-Verlag, Berlin, (2000).

[60] Klenk, H.; *"Cyano compounds, inorganic"*, in Gerhart, W. (Ed.) *Ullmann's Encyclopedia of Industrial Chemistry*, Vol. A8, pp. 159–163, VCH Publishers, New York, (1987).

[61] Hickman, D. A., Huff, M., Schmidt, L. D.; *"Alternative catalyst supports for HCN synthesis and NH_3 oxidation"*, Ind. Eng. Chem. Res. **32**, pp. 809–817 (1993).

[62] Andrussow, L.; *"Über die katalytische Oxidation von Ammoniak-Methan-Gemischen zu Blausäure"*, Angew,. Chem. **48**, pp. 593–604, 37 (1935).

[63] Andrussow, L.; *"Blausäuresynthese und die schnell verlaufenden katalytischen Prozesse in strömenden Gasen"*, Chem. Ing. Tech. **27**, pp. 469–472, 8/9 (1935).

[64] Kautter, C. T., Leitenberger, W.; *"Großtechnische Herstellung von Cyanwasserstoff nach Andrussow"*, Chem. Ing. Tech. **25**, pp. 679–768, 12 (1953).

[65] Endter, F.; *"Die technische Synthese von Cyanwasserstoff aus Methan und Ammoniak"*, Chem. Ing. Tech. **30**, pp. 305–310, 5 (1958).

[66] Wörz, O., personal communication (1999).

8   Gas/Liquid Microreactors

8.1   Gas/Liquid Contacting Principles and Classes of Miniaturized Contacting Equipment

Contacting of gases and liquids has been investigated in microreaction systems only recently [1–3]. After the development of design concepts, the first microsystems were realized and tested for their performance.

In principle, all generic flow arrangements for mixing of miscible fluids, derived in Chapter 3, can be applied for contacting of immiscible phases as well. To date, only two of them have been tested, namely "contacting of two substreams" and "injection of many substreams of two components" (see Chapter 3). Both approaches yield disperse gas/liquid systems composed of bubbles in a liquid medium. Another concept developed for gas/liquid contacting in microdevices, the utilization of falling films, has no analogue in mixing principles known for miscible fluids. This concept originally refers to the contacting of immiscible phases.

Flow Patterns of Disperse Gas/Liquid Systems
Generation of disperse systems is usually performed in flow arrangements based on a mixing tee configuration. These configurations are realized, e.g., by connection of a multichannel system to a mixing unit. After generation, the disperse phase is stabilized in these microchannels for a defined residence time. At present, three flow patterns have been reported for disperse phases in single microchannels (see Figure 8-1). At low gas space velocities either bubbly or segmented flow regimes are yielded which consist of alternately

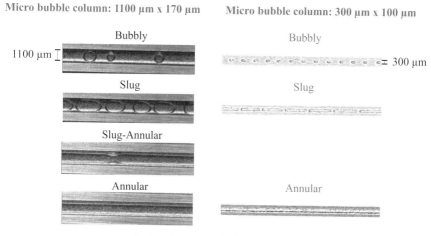

Micro bubble column: 1100 μm x 170 μm   Micro bubble column: 300 μm x 100 μm

Bubbly   Bubbly

1100 μm   300 μm

Slug   Slug

Slug-Annular

Annular   Annular

System: Isopropanol + Nitrogen

Fig. 8-1. Photographic images of two-phase flow corresponding to bubbly, slug, and annular flow patterns [5].

arranged gas bubbles and liquid films. Other terms for the segmented flow pattern commonly applied are slug or plug flow, see e.g. [4].

In the case of slug flow, the diameter of the gas bubbles approximately equals the channel diameter. A thin liquid film separates the bubbles from the channel walls. Moreover, thicker liquid films are found between the gas bubbles. Due to different velocities of gas and liquid phases, backmixing occurs in the disperse phases generated. At high gas superficial velocities, a transition from segmented to annular flow occurs. The latter phase is characterized by a gas core surrounded by a thin liquid hollow cylinder.

As a consequence of their correlation to the channel diameter, bubble sizes in the range of a few tens to a few hundred micrometers can be generated using state-of-the-art microfabrication techniques.

The flow patterns of multichannel arrays generally are more complex than single channels. In a micro bubble column, the formation of mixed flow patterns was observed, e.g. referring to slug/annular regimes. Most likely, this is caused by non-ideal flow equilibration and can be avoided in future versions of the microreactor with improved design of the mixing zone.

Specific Interfacial Areas of Gas/Liquid Microreactors and Conventional Equipment

For a rough estimation of the specific interfaces yielded for the disperse phases of the different flow patterns, assumptions about the thickness of the liquid layer between channel wall and gas bubble have to be made. Assuming this liquid layer to be only a few micrometers thick and measuring the bubble sizes and the length of the liquid plugs by means of light microscopy, specific interfacial areas up to 25,000 m^2/m^3 were detected (see Table 8-1). These values of the micro bubble columns are considerably larger than for specific interfaces yielded by standard equipment for gas/liquid contacting, as also revealed by

Table 8-1. Specific interfaces of 2-propanol/air in selected reactor types for gas/liquid contacting.

Single channels	Micro bubble bolumn	Flow pattern	Specific interfacial area (m^2/m^3)	
	1100 x 150 μm^2	Bubbly	1,700	
		Slug	8,700	
		Annular	8,600	
	300 x 100 μm^2	Bubbly	5,100	
		Slug	18,700	
		Annular	25,300	
Channel arrays in microreactors		Gas flow/ liquid flow (ml/h)	Average specific interfacial area (m^2/m^3)	Determined by sulfite oxidation (aqueous system)
	1100 x 170 μm^2	498/1000	5,100	–
	300 x 100 μm^2	300/20	16,600	9,800
	50 x 50 μm^2	–	–	14,800

the selection in Table 8-1. The best values of the conventional equipment are in the range of $2000 \ m^2/m^3$, e.g. achieved by mechanically stirred bubble columns.

As a consequence of the generation of mixed flow patterns in microchannel arrays, the corresponding interfacial areas are smaller than the highest values of the flow patterns of the single channels. Nevertheless, the former values still exceed the performance of conventional gas/liquid equipment, e.g. bubble columns and impinging jets, by an order of magnitude.

Falling Films

Generation of thin films is another way of achieving a high specific interfacial area. One principle commonly applied for the formation of these thin films is the contacting of liquid and gaseous phases in a falling film configuration [6]. So far, different reactor geometries have been used for falling film generation in conventional devices, e.g. including spherical, cylindrical and flat shapes. The latter concept, in particular, can be easily realized by means of microfabrication techniques. As one possible flat design, a sheet with multiple channels for flow guidance was developed (see Figure 8-2).

In this concept, the liquid phase is distributed on to the sheet, which is vertically oriented to the horizontal plane. Each substream enters a channel through a single orifice at the top of the sheet and flows downwards as a fluid film to a withdrawal zone at the bottom. While the flow direction of the liquid phase is determined by the action of gravity, the gaseous phase can be guided co- or countercurrently relative to the liquid phase.

The film thickness can be calculated by the following equation, derived from film theory [8, p.6–43].

$$\delta = \sqrt[3]{\frac{3 \cdot \dot{V} \cdot \eta}{\rho \cdot b \cdot n \cdot g}} \qquad (8.1)$$

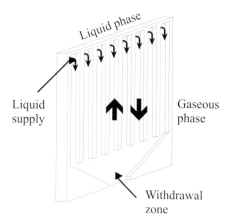

Fig. 8-2. Schematic of contacting liquid and gaseous phases in a falling film configuration [7].

Whereas the gravity constant, g, and physical parameters of the liquid phase, e.g. viscosity, η, and density, ρ, are fixed parameters for a given reaction system, the film thickness is determined for a given volume flow, $\dot{V}$, by width, b, and number, n, of the channels. Thus, the film thickness can be adapted to specific demands by variation of geometrical parameters of the devices.

In the following, three examples of gas/liquid microreactors utilizing these principles are given. Two refer to a mixing tee configuration which will be termed micro bubble column in the following. This term refers to similarities in construction to conventional bubble columns, although the flow regimes corresponding to the two types of reactors are different to some extent. Large-scale bubble columns are characterized by bubbly flow, whereas the miniaturized versions mainly display slug and annular flow patterns. The third example describes a falling film microreactor.

8.2 Contacting of Two Gas and Liquid Substreams in a Mixing Tee Configuration

8.2.1 Injection of One Gas and Liquid Substream

Researchers at the University of Newcastle and British Nuclear Fuels, Lancashire (U.K.) were among the first to describe flow patterns and issues for flow stability of gas/liquid phases in microchannels [2]. Since application orientation was an objective function for the work conducted, channel dimensions were chosen according to a compromise between throughput and stability of a two-phase flow. This was fulfilled by simple flow configurations similar to a mixing tee comprising 500 to 3000 µm wide and 200 to 800 µm deep channels.

Flow Patterns in Minichannels and Partially Overlapping Microchannels
In the framework of light-microscopy investigations, it was shown that a stable gas/liquid flow could be achieved in a 2000 µm wide channel, but restricted to a narrow flow range and with a limited stability. This flow is characterized by a gaseous core, connected to elongated segments in the center of the channel surrounded by liquid, hence being intermediate between slug and annular flow (see Figure 8-3). Similar experiments using parallel

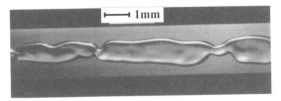

Fig. 8-3. Typical gas/liquid flow pattern obtained in a 2000 µm wide channel [2].

liquid/liquid systems reveal a significantly better flow stability over a wider range of operating conditions [2].

Reseachers at the Central Research Laboratories, Middlesex, and British Nuclear Fuels, Preston (U.K.), could extend the range of two-phase stability by using a two-channel system which only partially overlaps (see also Section 5.1.1) [3]. The authors performed an analysis of the benefits of miniaturization for gas absorption coupled to reactions covering a wide range of reaction kinetics. Conclusions, leading to different scenarios of transport limitations, were drawn on the basis of concentration profiles for the reactants absorbed and dissolved.

Ammonia Absorption

Examples of chemical reactions with practical and theoretical relevance have also been given [3]. For experimental investigations, CO_2 absorption into neutral or alkaline aqueous solutions was recommended as a slow to moderate process, depending on pH and presence of catalysts. However, no experimental investigation was carried out for this process, but rather for the fast NH_3 absorption into acidic aqueous solutions.

$$NH_3 \xrightarrow{\quad H_2O \quad} NH_4^{\oplus}$$

NH_3 absorption was carried out in a microcontactor device with 120 parallel microchannels of 14 mm length and 3000 μm^2 cross-section. Total liquid flow rates amounted to about 1 to 10 ml/h. The corresponding residence times ranged from 2 to 20 s. In contrast to the measurements discussed above, a fairly wide range of operating conditions yielded a stable two-phase flow. After passage through the reaction channel, the flows could be simply separated.

Depending on the NH_3 concentration, fast NH_3 absorption was recognized in these arrays of partially overlapping channels. In addition, the acid concentration was varied, resulting in different times until absorption was completed. These two types of concentration dependences indicate that, within the range of operating conditions, transport limitations exist for mass transfer both in solution and gas, even when using a microdevice.

8.2.2 Injection of Many Gas and Liquid Substreams into One Common Channel

In advance at contacting experiments in complex assembled microsystems, researchers of Institut für Mikrotechnik Mainz (Germany) performed a preliminary study on the dispersion of gas and liquid phases in an interdigital micromixer connected to a 3 mm polymeric tube [1]. Argon and water were chosen as a model system. It turned out that a stable bubbly flow in the tube could only be realized in the micromixer, whereas for a standard mixing tee only a plug flow with large Taylor bubbles resulted (see Figure 8-4). By addition of surfactants and increase of liquid viscosity, control over bubble size was exerted. Very narrow size distributions were achieved, e.g. for a special flow pattern, termed hexagon flow, corresponding to a mean size of 240±50 µm.

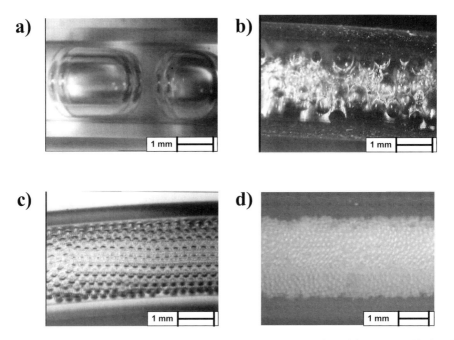

Fig. 8-4. Photographs of flow patterns of a pure water/argon disperse phase in a mixing tee (a) and the interdigital micromixer (b) as well as of glycerol, surfactant, water/argon phases in interdigital micromixers with a channel width of 40 μm (c) and 25 μm (d) [1].

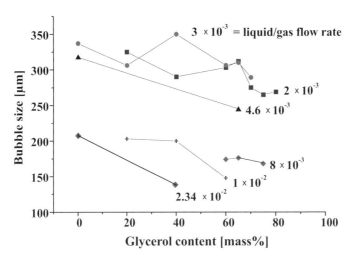

Fig. 8-5. Bubble sizes as a function of glycerol content in water, i.e. increasing viscosity, for various liquid/gas flow rates [1].

Further experiments focused on the influence of operating conditions on the bubble size. On increasing the liquid/gas flow rate, bubble sizes varying from 500 to 120 μm were achieved using the standard micromixing devices. It turned out that other important parameters are liquid viscosity and surface tension. By addition of glycerol, a decrease of bubble size was achieved as well, although less pronounced compared with the impact of fluid velocity (see Figure 8-5). The addition of a surfactant led to the formation of smaller and, in particular, more regular bubbles. In addition to operating conditions, the influence of geometric parameters was analyzed. On decreasing the width of the interdigital microchannels and of the discharge slit, both effects result in improved process performance.

Due to regular packing in a foam-like structure, the bubbles moved with a narrow residence time distribution. This was indirectly monitored by detection of the time dependence of light scattering of the disperse phases. In the case of uniformly sized bubbles, a nearly constant scattering intensity was found, whereas broader bubble size distributions resulted in a more or less periodic change in intensity. A sort of phase diagram was drawn yielding envelopes for uniform and non-uniform disperse phases as a function of the liquid/gas flow rate and the liquid viscosity.

8.2.3 Injection of Many Gas and Liquid Substreams into One Packed Channel

Packed Channel Microreactor
Researchers at the Massachusetts Institute of Technology (MIT), Cambridge (U.S.A.) developed a micro packed bed reactor to carry out heterogeneous catalyzed gas/liquid reactions (see Figure 8-6, left) [9]. The central element of this reactor is a reaction channel which is filled with catalytic particles for passage of gas/liquid dispersions. These dispersions are generated at the entrance region of the packed channel comprising a multichannel structure for each five gas and liquid inlet flows (see Figure 8-6, right). The parallel guided

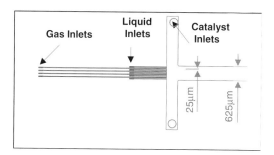

Fig. 8-6. Design of a micro packed bed reactor (left) and scanning electron micrograph of the multichannel mixing structure (right) [9].

235

single flows for gas and liquid phase are fed from two ports which are positioned at different distance from the inlet of the packed channel, thereby avoiding any crossover of the two phases.

The filling of the packed channel is usually achieved by pumping dispersions through an additional channel which is positioned vertically to the gas/liquid entrance region. By means of a second flow through the gas/liquid entrance region, part of the particles is redirected and finally fixed into the reaction channel. The particles are kept inside the reaction channel by a sieve-like structure at its end. Such a filling procedure was successfully demonstrated for polymer beads and catalytic powders. When using glass spheres, simple insertion within the channel is sufficient.

Microfabrication
The channel structure was realized by deep reactive ion etching. A further etching process was employed from the bottom side to manufacture ports for fluid feed and withdrawal. Finally, the structured silicon wafer was capped with a Pyrex glass plate and anodically bonded. By isolation of one interconnected silicon wafer, twelve single reactors of 15 mm x 35 mm size are yielded. Thereafter, the silicon/glass microreactor, compressed with one Viton gasket, is inserted in a housing with an aluminum top plate and stainless steel bottom plate. A heating cartridge is placed in a borehole within the top plate.

The gas and liquid inlet flow channels are 25 µm wide and approximately 300 µm deep separated by channel walls of similar width. The main reaction channel is 625 µm wide and 20 mm long. The vertical channel for catalyst filling amounts to 400 µm width and a volume of 4 µl.

Hydrogenation of α-Methylstyrene
Fluid flow visualization revealed a rapid mixing process between gas and liquid single flows at the entrance region of the reaction channel. As a test reaction, the hydrogenation of α-methylstyrene to cumene using carbon supported palladium catalysts was investigated. Depending on the ratio of the feed streams and the total flow rate, conversions between 20 % and 100 % were achieved at a temperature of 50 °C. Reaction rates as high as 0.01 mmol/min were determined, which agrees with kinetics reported in the literature.

Concluding Remarks
The micro packed bed reactor has a unique design and is one of the first concepts employed for carrying out reactions involving three phases–gas, liquid, and solid. The entrance flow region provides another design modification of a multilamination micromixing structure (see also Sections 3.7.1 to 3.7.4). In contrast to previous concepts, the single flows enter

the packed reactor in an absolutely parallel manner. A comparison of the impact of this type of flow guidance with that of similar structures for gas/liquid contacting, e.g. the microbubble column, discussed in Section 8.2.5, would give valuable insight. The use of packed catalyst particles in microchannels, although obviously not beneficial for gas phase reactions in microreactors (see Chapter 7), may be one useful approach for gas/liquid and liquid/liquid reactions including solid surfaces.

8.2.4 Injection of Many Gas Substreams into One Liquid Channel with Catalytic Walls

Modeling of Flat Plate Microreactor
A theoretical parameter study of gas/liquid catalytic reactions using a flat plate microreactor (see Figure 8-7) was performed at the University College London (U.K.) [10]. The goal of this study was to show the dependence of conversion of this type of processes on basic influence quantities such as velocity, critical reactor dimension, reactant diffusivity, and catalyst loading. As model reaction, hydrogenation at catalytic surfaces was employed. The flat plate reactor consisted of a porous wall where hydrogen entered the reaction channel, referred to as diffusion wall, and a catalytic wall opposite to it. For modeling, a commercial finite volume-based CFD code (CFDRC) was used. Since it was assumed that the concentration of hydrogen does not exceed its solubility concentration (here 5 wt.-%), only two phases, namely solid and liquid, need to be considered when modeling the reaction.

Concentration Profiles During Reaction
For a reaction mixture, a flow velocity of 0.01 m/s and a distance of 100 μm between diffusion and catalytic walls, axial and radial product concentration profiles were monitored (see Figure 8-8). Only close to the channel exit was, the onset of reaction observed. The

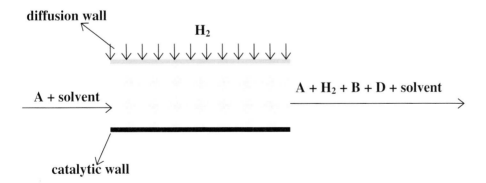

Fig. 8-7. Schematic of a flat plate microreactor for performing liquid reactions with dissolved gas [10].

237

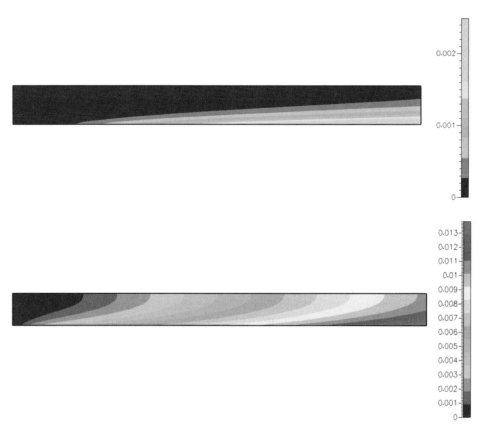

Fig. 8-8. Concentration profiles within a reaction channel of a flat plate microreactor for an idealized liquid reaction with dissolved gas [10]. Profile assuming a flow velocity of 0.01 m/s (top), profile assuming a flow velocity of 0.001 m/s (bottom).

decrease in velocity by a factor of ten resulted in an increasing concentration of the hydrogenated product all over the channel length. The radial concentration profiles are less steep than for a reaction carried out at higher velocity.

A decrease in wall distance results, as to be expected, in a strong increase in conversion. For instance, for a 30 μm microchannel only 25 % conversion was observed, whereas a further reduction in distance from the hydrogen inlet to the catalytic surface to 10 μm leads to 100 % conversion. A similar effect is observed on systematically reducing the reaction mixture velocity and the residence time. A nearly linear increase in conversion as a function of the diffusion coefficient was calculated. In contrast, a higher catalyst loading, equivalent to a larger reaction rate, had only a minor effect on conversion enhancement.

Concluding Remarks

For a simple microreactor design, the work conducted revealed some interesting fundamental aspects for further reactor design. The applicability to real aspects of reaction engineering concerning microreactors will be enlarged if future studies take into account more complex types of processing, e.g. including heat release and transfer as well as disperse gas/liquid phases.

8.2.5 Injection of Many Gas and Liquid Substreams into Multiple Channels

Researchers at the Institut für Mikrotechnik Mainz (Germany) realized a micro bubble column [6, 10, 11] using multilamination as the mixing principle as already described in Chapter 6. The research conducted aimed at a proof of the feasibility of using gas/liquid reactions established for basic characterization and transfer of these results into application. As an industrially relevant example, the direct fluorination of aromatic compounds using elemental fluorine was chosen (see Section 8.3).

Construction of the Micro Bubble Column

The central component of the micro bubble column is a static micromixer (see Figure 8-9) which is mechanically connected to a reaction platelet. An adjustment of mixer and heat exchanger with a precision of a few micrometers is provided by a two-piece housing, manufactured employing ultraprecision techniques. In this housing, a micro heat exchanger and fluid inlets and outlets for the mixing and reaction units are integrated.

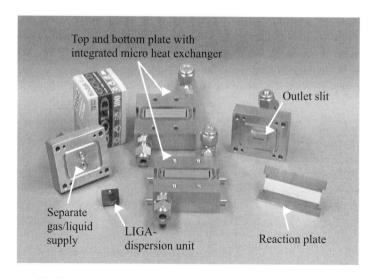

Fig. 8-9. Components of the micro bubble column [7].

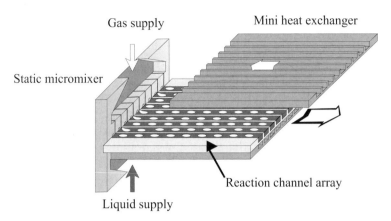

Fig. 8-10. Schematic of a static micromixer with an interdigital channel configuration with gas and liquid feed channels of different size [7].

The micromixer was based on a flow configuration first applied for multilamination of miscible fluids [11]. Despite this structural analogy, two distinct differences from the original design have to be mentioned. First, the width of the gas feed channels is much smaller than that of the liquid feed channels in order to guarantee flow equipartition by applying a large pressure barrier. Second, instead of multilaminating gas and liquid streams in one common discharge slit, each gas/liquid bilayer stream is fed into a separate reaction channel (see Figure 8-10).

For the microchannel array, formed by parallelization of many of these channels, thermal control is provided by highly efficient heat transport to a heat transfer fluid flowing in countercurrent mode. This fluid flows in minichannel arrays embedding the reaction channel array from both sides.

Design Criteria
The design criteria for the microbubble column were oriented on assumptions concerning fluorine reactivity and reaction enthalpy data. Due to the absence of reliable kinetic data, the reaction rates could only be estimated from solubility data of fluorine. The latter were corrected by an enhancement factor, due to chemical reaction, for mass transport from gas into the liquid, the so-called Hatta number.

All these assumptions on reactivity had to be complemented by estimations regarding hydrodynamics, predicting the flow regime. Based on investigations also in monoliths with square or circular capillaries [12], a slug flow pattern was expected to be establish for a two-phase flow in the microchannels [12, 13]. The correctness of this estimation was later verified by experiments.

From the measurements in the monoliths, it was further assumed that the diameters of the gas bubbles approximately correspond to the channel width. In addition, these investigations showed that the ratio of bubble length to diameter is of the order of one to four.

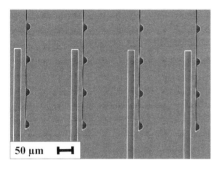 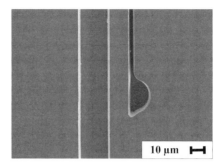

Fig. 8-11. Scanning electron micrographs of the interdigitated gas and liquid feed channels of the micromixer [7].

Finally, order-of-magnitude values were available for the thickness of the liquid films separating the bubbles from each other and the channel wall. From these data, large specific interfacial areas of several 10,000 m^2/m^3 were calculated for two-phase systems in square or rectangular microchannels with a hydraulic diameter of a few 10 to 100 µm (see Table 8-1).

The criteria based on conversion and heat release, as mentioned above, were met by rectangular reaction microchannels designed with a width, depth and length of 300 µm, 150 µm and 50 mm, respectively. In addition, microchannels even smaller in cross-section were employed, amounting to 50 µm x 50 µm. The specific interfacial areas calculated for these two channels were 14,000 and 40,000 m²/m³, respectively.

The parallel operation of a high number of channels demands flow equipartition of gas and liquid streams. For this purpose, a pressure barrier was utilized. Pressure loss calculations showed that the width of the gas mixing channels has to be as small as 5 µm, whereas the width of the liquid mixing channels should not exceed 25 µm (Figure 8-11).

Microfabrication
Micromixer platelets with various gas and liquid channel widths were realized in nickel using a UV-LIGA process. Reaction platelets were fabricated with varying channel dimensions as well. Microchannels of larger cross-section of 300 µm x 150 µm were obtained by means of wet chemical etching using a high-alloy stainless steel. Smaller reaction channels of 50 µm x 50 µm size were achieved in nickel by applying a UV-LIGA process. The housing was made of high-alloy stainless steel by precision engineering and micro electro discharge machining using a micro die sinking process.

Specific Interfacial Areas
The theoretical calculations of the specific interfacial area were made using experimental data gathered by the widely used sulfite oxidation method [14]. For the two-phase flow in the microchannels of large cross-section, interfacial areas of up to 9000 m^2/m^3 were meas-

ured which are of the same order as the calculated value of 14,000 m²/m³. Thus, it can be concluded that bubbles with a diameter and length of the order of 300 μm and 1200 μm, respectively, were generated.

In addition, light microscopy observation of gaseous and liquid flows in each channel allowed the control of flow equipartition. Thereby, the co-ocurrence of more than one flow pattern in the microchannel array was identified. For instance, a bubbly/slug/annular regime was observed. For this reason, the mean specific interfacial areas of the array are smaller than the best values for the single channels. Although this non-ideal flow equipartition demands further design improvements, it has to be pointed out that the quality of flow partition is better as than that of conventional gas/liquid monoliths.

Carbon Dioxide Absorption

The high specific interfacial areas determined by light microscopy and sulfite oxidation (see Table 8-1) suggest a superior performance of the micro bubble columns over conventional equipment and multi-purpose microdevices. Accordingly, a comparison of mass transfer efficiency was performed including testing of different special-type gas/liquid microreactors, such as three micro bubble columns [15], multi-purpose contacting microdevices, such as two micromixers [11, 16, 17], and simple conventional equipment, such as one mixing tee (see Figure 8-12). The mixing tee, having an internal diameter of 1 mm, was used as a model system for capillaries of similar dimensions, e.g. as applied in monoliths. In addition, a benchmarking to the efficiency of packed columns in terms of space–time yields was performed (see Table 8-2 in Section 8.3).

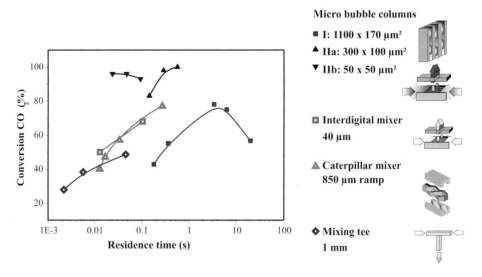

Fig. 8-12. Comparison of reactor performance of special-type gas/liquid and multipurpose microdevices and also one mixing tee as evidenced by the conversion of carbon dioxide as a function of residence time [15].

$$CO_2 \xrightarrow{\;\;OH^{\ominus}\;\;} HCO_3^{\ominus}$$

For comparison of mass transfer efficiency, carbon dioxide absorption was used as a model reaction [14, 15, 18]. In Figure 8-12, the conversions of carbon dioxide achieved are shown as a function of the mean residence time of both fluid phases in the different reactors. In micro bubble columns IIa and IIb (small-sized) nearly 100 % conversion of carbon dioxide is achieved, even for residence times below 0.1 s in the case of device IIb. For near equivalent residence times, e.g. 0.1 s, mass transfer of carbon dioxide is more efficient in the smaller microchannels of micro bubble column IIb as compared with device IIa. The increase of carbon dioxide conversion of micro bubble column IIa by a factor of 2.5 compared with device I at a residence time of about 0.1 s is due to the same origin. Finally, the mixing tee comprises channels of the largest dimensions, hence exhibiting the lowest performance.

For all devices, the efficiency reaches a plateau or even passes a maximum as residence time, or flow energy, is varied. Similar findings are known from measurements of average droplet size as a function of volume flow when contacting liquid/liquid streams in an interdigital micromixer [19]. In the low volume flow range, a decrease of droplet size is caused by an increase in flow energy. However, increasing volume flow further results in reduced performance due to the decrease of residence time. As a consequence of this antagonism between flow energy and residence time, performance is characterized by the presence of a maximum.

In the case of gas/liquid contacting, these simple mechanistic correlations are superposed by the change of flow patterns with increasing volume flow. The overall effect, however, is the same, namely an increase in specific interfacial area and, correspondingly, an increase in carbon dioxide conversion.

Concluding Remarks
Apart from simply phenomenologically describing the flow patterns of gas/liquid contacting in microchannels, as partially performed in previous work [2, 20], the characterization of the micro bubble columns provides for the first time quantitative information on hydrodynamics, with regard to stability of flow patterns and their respective interfacial areas. For the first time, flow pattern maps of microchannels are given and interfacial areas are contrasted with conventional systems. In particular, the impact of the scale of the microchannels on the hydrodynamics was evidenced. These fundamental studies were completed by analyzing the mass transfer properties using the carbon dioxide absorption. Thereby, the performance of special-type microreactors was compared with that of multi-purpose tools, e.g. micromixers. Finally, the fluorination of toluene was performed in the micro bubble column as an example of use with industrial relevance (see Section 8.3)

8.3 Generation of Thin Films in a Falling Film Microreactor

A falling film microreactor was developed at the Institut für Mikrotechnik Mainz (IMM, Germany) simultaneously with the manufacture of the micro bubble column. The corresponding investigations focused on similar topics, namely to prove the feasibility of miniaturization of a macroscopically known gas/liquid contacting principle and to test reactor performance by direct fluorination of aromatic compounds.

Construction of the Falling Film Microreactor
The falling film microreactor consists of four components, shown in Figure 8-13, namely a bottom plate with an integrated heat exchanger, a reaction plate, a contact zone mask and a top plate with a gas chamber. The liquid flows as a falling film in the reaction plate, whereas the reactant gas is introduced in a rectangular-shaped gas chamber in the top plate via diffusor nozzles. Thermal control is provided by heat exchange between the reaction plate and underlying heat exchanger channels in the bottom plate.

The main element of the microreactor is the reaction plate comprising a microchannel array. Supply and withdrawal of liquid reactants to the microchannels are achieved through numerous small orifices within the plate connected to two boreholes, referring to inlet and outlet zones, via two large slits (see Figure 8-14).

For distributing a heat transfer medium among the heat exchanger channels in the bottom plate, the same manifolding principle is used. The width and depth of these channels match the dimensions of the reaction channels. Their length, however, has to be shorter due to a reduction of space available owing to the need to insert additional boreholes for feed and withdrawal flows. In order to compensate for this difference in length for reaction and cooling channels, a so-called contact zone mask covered the entrance and exit region of the reaction channel array.

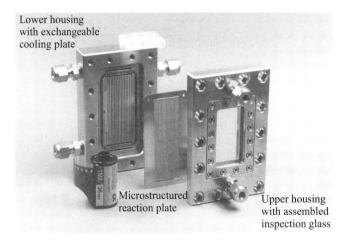

Fig. 8-13. Components of falling film microreactor [7].

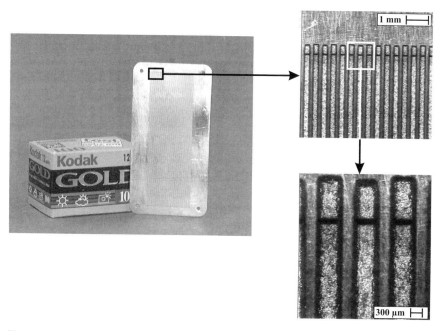

Fig. 8-14. Supply orifices in the reaction plate connected via one large slit [7].

Thereby excluding any contact of the entering flow to the reactant gas, the reaction is restricted to the heat exchange zone and, hence, held under favorable thermal control. By precise adjustment of the reaction plate on top of the bottom plate, each reaction channel is positioned exactly above a single heat exchanger channel.

Microfabrication

The width and depth of the reaction and heat exchange channels were set to 300 µm and 100 µm, respectively, whereas the length amounted to several tens of millimeters. Microfabrication of these channels was performed by means of micro electro discharge machining (µ-EDM) and wet etching of thin metal foils. In the case of µ-EDM, channels were realized by a die sinking process using electrodes, fabricated by thin-wire microerosion, with inverse shapes of the microstructures.

Orifices and slits for feed and withdrawal flows, having heights of 80 µm and 1 mm, respectively, were also generated by die sinking. The outer contours of the contact zone mask were cut using laser machining, while the fabrication of the gas chamber was performed by die sinking.

Design criteria of the falling film microreactor were based on the data and assumptions on heat release and chemical reactivity already applied for the micro bubble column. With respect to hydrodynamics, preliminary experimental investigations focused on the determination of film thicknesses at varying operating parameters.

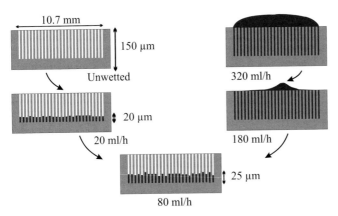

Fig. 8-15. Graphic representation of profile measurements of the reaction plate with different operation modes (unwetted, partially filled at different stationary volume flows, partially and totally flooded) [7].

Determination of Film Thickness

A non-contact optical method, using a Microfocus-UBM® instrument, proved to be successful for this purpose. This technique provides precise topological information by an autofocus system based on total reflection of the incoming light beam on a surface. Data on film thickness were gained by subtracting the vertical positions determined by light reflection at the channel's bottom and liquid surface. All experiments were carried out with dead gas load, i.e. the reaction plate being uncovered by the top plate. In Figure 8-15, results for film thickness measurements, corresponding to different operating conditions, were translated in a graphic representation which reveals the cross-section of the reaction plate including the liquid film.

This collection provides a profile of the unwetted reaction plate and two profiles of only partially filled microstructures at different stationary volume flows. Additionally, two measurements referring to operation modes characterized by partially and totally flooded reaction plates are given. Comparing the first four plots, an increase of film thickness with volume flow is evident. Film thicknesses measured were in the range of the theoretical values derived from equation (3). However, the number of data was insufficient to prove the validity of this equation for measurements of micrometer-sized films as well.

Flow Equipartition

As for the micro bubble column, the quality of flow equipartition was examined, e.g. by means of photographic techniques and tracer experiments. For imaging of streaming fluids, the starting wetting behavior of the reaction plate with dead gas load, i.e. being uncovered by the top plate, was characterized. It could be proved that these results are representative for stationary flow conditions as well.

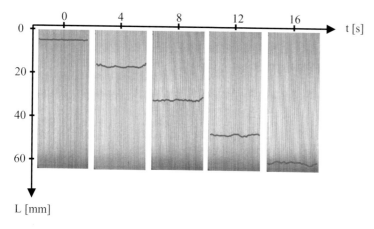

Fig. 8-16. Streaming fluid front (acetonitrile) in an open channel system at different times. Flow profiles were traced graphically in order to enhance the contrast of the photographs [7].

Figure 8-16 shows the time development of a flowing fluid front, by revealing five flow profiles corresponding to time intervals each separated by 4 s. As solvent, acetonitrile with a volume flow of 20 ml/h was chosen.

A cumulative plot of the residence times of all flows in the single microchannels proves very uniform residence time distributions. For 90 % of these flows, the average residence time amounts to 17.5 ± 0.5 s (see Figure 8-17).

Direct Fluorination of Toluene: Set-up and Experimental Protocol
As an example of use with industrial relevance, the direct fluorination of toluene, dissolved in acetonitrile, with elemental fluorine was selected. Currently, this process cannot be per-

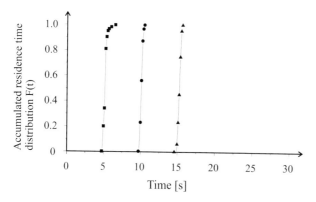

Fig. 8-17. Cumulative frequency distribution at different times [7].

formed on an industrial scale due to limitations of mass and heat transfer as well as for reasons of process safety. Instead, multi-step processes utilizing nucleophilic substitution are employed for large-scale production of fluorinated aromatic compounds, namely the Schiemann [21, 22] and Halex [22, 23] processes.

A set of experiments was carried out at the Institut für Angewandte Chemie, Berlin (Germany) and at the Forschungszentrum Karlsruhe (Germany) using 10 % F_2 in N_2 as gas reactant stream, the reactor performance of the falling film microreactor and the micro bubble column was evaluated. These results were compared with fluorination data achieved using a laboratory bubble column, a macroscopic gas/liquid contactor widely used for investigations in chemistry. This device has been used in the vast majority of relevant scientific investigations described in this field [24–26]. It has to be stated that much more elaborate conventional gas/liquid devices exist. However, the fact is that there is no example for the use of such a device for direct fluorination. Most likely, this is strongly prohibited for reasons of process safety.

$$\text{(benzene ring)} \xrightarrow[\text{CH}_3\text{CN}]{F_2} \text{(benzene ring with F)} + \text{(benzene ring with F)}$$

During experimental investigations, considerable efforts had to be made to establish suitable analytical methods for gathering reliable data. It turned out that using high-performance liquid chromatography (HPLC) quantitative data on toluene consumption and generation of monofluorinated products can be obtained. Due to fluorine-to-toluene ratios < 1 applied in most experiments, it would have been preferable to refer conversion, yield, and selectivity to the understoichiometric component, i.e. fluorine. However, this was not possible for experimental reasons and, consequently, the corresponding definitions had to be based on toluene.

Conversion and Selectivity
In a first set of experiments, the molar ratio of fluorine to toluene was varied, ranging from 40 to 100 % [7]. Although, at the time of performing the experiments very limited information about the proper setting of operating parameters for the microreactors was available, already the very first attempts were able to demonstrate good reactor performance. For instance, the corresponding conversions, based on toluene, were of the order of 30 to 50 % for the falling film microreactor and 12 to 28 % for the micro bubble column. These data exceeded those of the laboratory bubble column, in particular in the case of the falling film microreactor. Correcting these data according to the understochiometric ratio applied reveals that in some experiments about 70 % of all toluene molecules, which potentially could react, were converted.

Even more remarkable, relatively high selectivities of about 36 to 49 % were achieved using the microreactors. The same holds for the yields derived which, with the exception of one measurement, exceeded those of the laboratory bubble column. Comparing the best data obtained for macro- and microscopic devices, the yield of 20 % measured for the

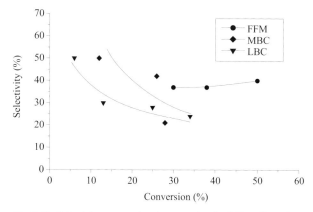

Fig. 8-18. Selectivity–conversion diagrams for the falling film microreactor (FFM) and micro bubble column (MBC) compared with the benchmark, the laboratory bubble column (LBC) [7].

falling film microreactor is 2.5 times larger than the corresponding benchmark of the laboratory bubble column.

Operation using highly concentrated fluorine gas mixtures, e.g. 50 % F_2 in N_2, resulted in yields of monofluorinated toluenes (*para* and *ortho*) of up to 28 %. This value is comparable to the total yield of the multi-step syntheses of the Schiemann and Halex processes. In additional fluorination experiments, even pure toluene was contacted with these highly concentrated gas mixtures. Although the corresponding yields were lower than when using acetonitrile solutions, the safe performance of such an experimental protocol in microreactors, otherwise being of notable explosive nature, is a success on its own and paves the way for further investigations.

Another graphical representation often used to characterize the performance of reactors is a selectivity–conversion diagram (see Figure 8-18). Although the data set available for this diagram was small, there is a clear indication of an improved performance of the microreactors, in particular concerning the falling film reactor, compared with the laboratory bubble column.

Space–Time Yields

As a measure of potential economical benefit, the space–time yields of the microreactors were compared with the benchmark (see Figure 8-19) [27]. To yield a definition focusing on reaction engineering aspects, the microreactor volume taken into account was based solely on the microchannel volume, thus omitting the housing volume. The space–time yields derived in this manner reveal a performance of the microreactors that is better by orders of magnitude. Even including the currently large housing volume, performance still remains improved. By optimizing microfabrication and interconnection strategies, it is very likely that the housing volume can be substantially reduced when developing the present demonstrators of gas/liquid microreactors towards future generation prototypes.

a)

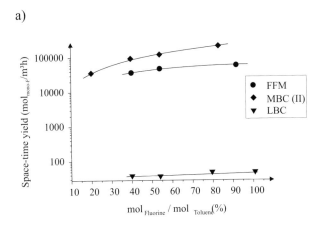

b)

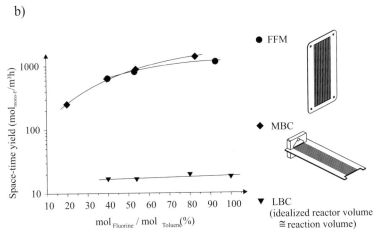

Fig. 8-19. Space–time yields of the falling film microreactor (FFM) and micro bubble column (MBC) compared with the benchmark, the laboratory bubble column (LBC) [27].

Substitution Pattern

In addition to the analysis of reactor performance, experiments were performed to eluci-date the reaction mechanism [7]. Monosubstitution of toluene by fluorine theoretically can lead to three different isomers, their relative ratio being dependent on the mechanism of reaction. A substitution pattern of 5 : 1 :3 for *ortho*, *meta* and *para* isomers, respectively, was determined from the peak areas of the GC plot (see Figure 8-20). A similar distribution of products was determined using IR spectroscopic methods based on attenuated total re-flection (ATR) techniques. This pattern is consistent with an electrophilic substitution mecha-

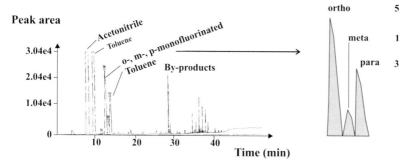

Fig. 8-20. Typical product spectrum of direct fluorination using acetonitrile achieved in the falling film microreactor [7].

nism that has been proposed for this reaction by mechanistic studies using highly diluted solutions [25].

Benchmarking to Industrial Processes
A semi-quantitative analysis of GC and FTIR results confirmed the information gathered from HPLC, hence indicating a significant generation of monofluorinated products relative to the undesired by-products as well [27]. The best yields determined by these methods were up to 28 %. With respect to the substitution pattern, about 11 % of *ortho* product and 7 % of *para* product are formed. The Balz–Schiemann process, commonly applied in industrial processes, yields about 25 % of only one isomer. Compared with the microreactors, the total yield of this process is higher and the expenditure for product separation is lower. However, the microreactors use a one-step process, with potentially less waste generation as well as expenditure of energy and raw materials, whereas the Balz–Schiemann route is a multi-step process.

Carbon Dioxide Absorption
Similarly to the characterization of the micro bubble column (see Section 8.2.5), absorption of CO_2 in NaOH was chosen as a model reaction to characterize the efficiency of mass transfer in the falling film microreactor. In all experiments the molar ratio of CO_2 to NaOH was kept constant at 0.4, i.e. referring to an excess of 20 % of liquid reactant. Liquid reactant concentration and film thickness being set, gas flow and concentration had to be adapted according to this ratio of reactants.

$$CO_2 \xrightarrow{\;OH^{\ominus}\;} HCO_3^{\ominus}$$

A first set of experiments referred to an analysis of the impact of reactant concentration and gas flow direction, liquid flow and film thickness being fixed. A liquid flow of 50 ml/h was

251

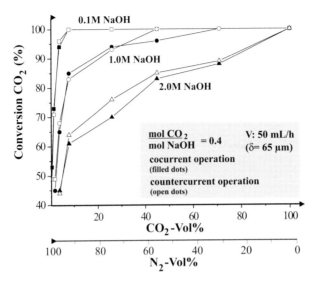

Fig. 8-21. Conversion of CO_2 as a function of reactant concentration and flow direction of the gaseous phase for the first experimental set (total liquid flow 50 ml/h, 64 single flows of 65 μm film thickness, ratio of CO_2 to NaOH 0.4) [15].

chosen resulting, for a falling film microreactor, in 64 microchannels of 300 μm width and 100 μm depth, in a film thickness of approximately 65 μm, a specific surface area of about 15,000 m^2/m^3, and a liquid residence time of about 6 s according to Nusselt's theory [28, 29].

In Figure 8-21, CO_2 conversions obtained for the first experimental set are shown. For fixed NaOH concentration, decreasing CO_2 concentration in the gas phase and, thus, increasing gas velocities result in lower CO_2 conversions. The same behavior is observed with increasing NaOH concentration in the medium to low range of CO_2 concentrations. The gas flow direction, either co- or countercurrent operation, has no significant effect on CO_2 removal from the gaseous feed with the 0.1 M and 1.0 M NaOH solutions.

With regard to the determination of optimum processing conditions, it has to be mentioned that complete CO_2 absorption can be achieved for both low and highly concentrated liquid flows. This is achievable by using CO_2 gas streams of relatively low velocity and, hence, large gas flow residence times. The finding of complete absorption for all NaOH concentrations investigated shows that both liquid and gaseous transport can be set sufficiently fast to permit reaction for the given ratio of CO_2 to NaOH. The decrease of CO_2 conversion on decreasing the CO_2 concentration in the gas phase is possibly caused by liquid-side hindrance in the range of high CO_2 concentrations, changing to gas-side hindrance at low CO_2 concentrations. Such an interplay of transport hindrance has been reported in a number of cases for carbon dioxide absorption, dependent on the operating conditions [14, 30].

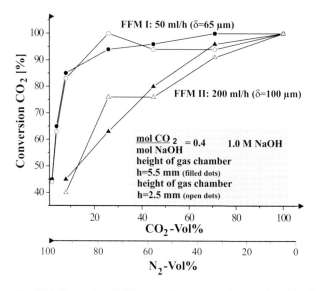

Fig. 8-22. Conversion of CO_2 depending on gas velocity and liquid volume flow in a falling film microreactor utilizing reaction platelets of various size and, consequently, variable volume flow [15].

Furthermore, a second experimental set was carried out using a modified gas chamber the height of which was reduced from 5.5 mm to 2.5 mm (see Figure 8-22). Thereby, the gas velocity was increased by a factor of 2.2 for fixed CO_2 concentration. In addition to these modifications of gas-side design, the effect of increasing liquid volume flow by means of generating thicker films was investigated. For this, a reaction plate with 15 reaction channels, 1.2 mm wide and 400 μm deep, was manufactured, allowing experiments to be performed at higher liquid load. Thereby, a volume flow of 200 ml/h resulting in a film thickness of 100 μm could be utilized. Compared with the previous design with 64 channels operated at 50 ml/h and a film thickness of 65 μm, the specific surface area is decreased from 15,000 m^2/m^3 to 10,000 m^2/m^3.

The results of the second experiment set, using an NaOH concentration of 1.0 M, are shown in Figure 8-22. Even in the range of low CO_2 concentrations, from 2 vol.-% to 8 vol.-% CO_2 with corresponding gas velocities of 4.3 to 9.4 cm/s, there is no evidence of a notable influence of doubling of the gas velocity, with either a 50 ml/h or 200 ml/h liquid flow. The increase in gas velocity, however, has one pronounced effect, namely to induce higher deviations of the measured values, probably caused by varying flow patterns.

The influence of increasing film thickness (from 65 μm to 100 μm) is even more difficult to interpret since, in addition, two other parameters are changed simultaneously, namely liquid residence time and gas velocity. Since all these single influence quantities potentially lead to lower CO_2 conversion, it was not surprising that the experimental results showed as an overall result a dramatic decrease of CO_2 removal.

In Table 8-2 the space–time yields of the falling film microreactor, the micro bubble column, and a conventional packed column are listed for similar operating conditions. The space–time yields are referred to the reaction volume in the microchannels. For a discussion on the relation of this figure with respect to corresponding definitions based on the total system size, see [27]. It turned out that the space–time yields of both microreactors exceed the performance of the standard equipment by orders of magnitude. Moreover, the performance of the microreactors can be increased by increasing the volume flow or liquid reactant concentration.

Table 8-2. Space–time yields of carbon dioxide absorption for comparable operating conditions when using the falling film microreactor, a micro bubble column, and a conventional packed column [31].

Reactor type	Conc. NaOH (mol/l)	Conc. CO_2 (vol.-%)	Molar ratio CO_2/NaOH	Conversion CO_2 (%)	Space–time yield (mol/m^3 s)
Packed column	1.2	12.5	0.41	87	0.61
Packed column	2.0	15.5	0.43	93	0.81
Falling film microreactor	1.0 (50 ml/h, 65 µm)	8.0	0.40	85	56.1
Falling film microreactor	2.0 (50 ml/h, 65 µm)	8.0	0.40	61	83.3
Falling film microreactor	1.0 (200 ml/h, 100 µm)	8.0	0.40	45	83.7
Micro bubble column (50x50 µm^2)	2.0 (10 ml/h)	8.0 (2400 ml/h)	0.33	100	227
Micro bubble column (50x50 µm^2)	2.0 (50 ml/h)	8.0 (12,000 ml/h)	0.33	72	816

Concluding Remarks

The first results, presented in Sections 8.2 and 8.3, demonstrate clearly the potential for improvements of gas/liquid processes by enhancing mass and heat transfer. Therefore, the falling film concept turned out to be a versatile approach characterized by very uniform residence time distributions and reliable generation of specific interfacial areas. The latter values are higher than any other interfacial area reported to date. Consequently, mass transfer in the falling film microreactor, as evidenced by carbon dioxide absorption, exceeds the performance of today's best gas/liquid contactors, e.g. impinging jets.

The results of the fluorination of toluene indicate that a possible implementation in industry is not far from being reality. By using reaction platelets of increased volume throughput, as recently developed at IMM, a further step for reaching this target was made. In addition, a commercial small-series fabrication of this microreactor was started just recently. In the near future, a reactor model will be developed, thereby further increasing the range of applicability and allowing the prediction of the behavior of a microreactor concerning new reactions.

8.4 References

[1] Hessel, V., Ehrfeld, W., Golbig, K., Haverkamp, V., Löwe, H., Richter, T.; *"Gas/liquid dispersion processes in micromixers: the hexagon flow"*, in Ehrfeld, W., Rinard, I. H., Wegeng, R. S. (Eds.) *Process Miniaturization: 2nd International Conference on Microreaction Technology, IMRET 2; Topical Conference Preprints*, pp. 259–266, AIChE, New Orleans, USA, (1998).

[2] Burns, J. R., Ramshaw, C., Bull, A. J., Harston, P.; *"Development of a microreactor for chemical production"*, in Ehrfeld, W. (Ed.) *Microreaction Technology, Proceedings of the 1st International Conference on Microreaction Technology; IMRET 1*, pp. 127–133, Springer-Verlag, Berlin, (1997).

[3] Shaw, J., Turner, C., Miller, B., Harper, M.; *"Reaction and transport coupling for liquid and liquid/gas micro-reactor systems"*, in Ehrfeld, W., Rinard, I. H., Wegeng, R. S. (Eds.) *Process Miniaturization: 2nd International Conference on Microreaction Technology, IMRET 2; Topical Conference Preprints*, pp. 176–180, AIChE, New Orleans, USA, (1998).

[4] Triplett, K. A., Ghiaasiaan, S. M., Abdel-Khalik, S. I., Sadowski, S. L.; Int. J. Multiphase Flow **25,** pp. 377–394 (1999).

[5] Results of IMM, unpublished

[6] Astarita, G.; *Mass Transfer with Chemical Reaction*, p. 94, Elsevier, Amsterdam (1967).

[7] Hessel, V., Ehrfeld, W., Golbig, K., Haverkamp, V., Löwe, H., Storz, M., Wille, C., Guber, A., Jänisch, K., Baerns, M.; *"Gas/liquid microreactors for direct fluorination of aromatic compounds using elemental fluorine"*, in Ehrfeld, W. (Ed.) *Microreaction Technology: 3rd International Conference on Microreaction Technology, Proceedings of IMRET 3*, pp. 526–540, Springer-Verlag, Berlin, (2000).

[8] Perry, R. H., Green, D. W.; *Perry's Chemical Engineers' Handbook,* 7th ed; McGraw-Hill, New York (1997).

[9] Losey, M. W., Schmidt, M. A., Jensen, K. F.; *"A micro packed-bed reactor for chemical synthesis"*, in Ehrfeld, W. (Ed.) *Microreaction Technology: 3rd International Conference on Microreaction Technology, Proceedings of IMRET 3*, pp. 277–286, Springer-Verlag, Berlin, (2000).

[10] Angeli, P., Gobby, D., Gavriilidis, A.; *"Modelling of gas-liquid catalytic reactions in microchannels"*, in Ehrfeld, W. (Ed.) *Microreaction Technology: 3rd International Conference on Microreaction Technology, Proceedings of IMRET 3*, pp. 253–259, Springer-Verlag, Berlin, (2000).

[11] Ehrfeld, W., Golbig, K., Hessel, V., Löwe, H., Richter, T.; *"Characterization of mixing in micromixers by a test reaction: single mixing units and mixer arrays"*, Ind. Eng. Chem. Res. **38,** pp. 1075–1082, 3 (1999).

[12] Thulasidas, T. C., Abraham, M. A., Cerro, R. L.; Chem. Eng. Sci. **50,** pp. 183–199, 2 (1995).

[13] Hatziantoniou, V., Andersson, B.; Ind. Eng. Chem. Fundam. **23,** pp. 82–88 (1984).

[14] Deckwer, W.-D.; *Reaktionstechnik in Blasensäulen,* Otto Salle Verlag, Frankfurt/M (1984).

[15] Hessel, V., Ehrfeld, W., Herweck, T., Haverkamp, V., Löwe, H., Schiewe, J., Wille, C., Kern, T., Lutz, N.; *"Gas/liquid microreactors: hydrodynamics and mass transfer"*, in Proceedings of the "4th International Conference on Microreaction Technology, IMRET 4", pp. 174–186; 5–9 March, 2000; Atlanta, USA.

[16] Löwe, H., Ehrfeld, W., Hessel, V., Richter, T., Schiewe, J.; *"Micromixing technology"*, in Proceedings of the "4th International Conference on Microreaction Technology, IMRET 4", pp. 31–47; 5–9 March, 2000; Atlanta, USA.

[17] Ehrfeld, W., Hessel, V., Löwe, H.; *"Extending the knowledge base in microfabrication towards thermical engineering and fluid dynamic simulation"*, in Proceedings of the "4th International Conference on Microreaction Technology, IMRET 4", pp. 3–22; 5–9 March, 2000; Atlanta, USA.

[18] Fleischer, C., Becker, S., Eigenberger, G.; *"Transient hydrodynamics, mass transfer, and reaction in bubble columns: CO_2 absorption into NaOH solutions"*, Trans. Inst. Chem. Eng. **73-A,** 8 (1995) .

[19] Schiewe, J., Ehrfeld, W., Haverkamp, V., Hessel, V., Löwe, H., Wille, C., Altvater, M., Rietz, R., Neubert, R.; *"Micromixer based formation of emulsions and creams"*, in Proceedings of the "4th International Conference on Microreaction Technology, IMRET 4", 5–9 March, 2000; Atlanta, USA.

[20] Burns, J. R., Ramshaw, C., Harston, P.; *"Development of a microreactor for chemical production"*, in Ehrfeld, W., Rinard, I. H., Wegeng, R. S. (Eds.) *Process Miniaturization: 2nd International Conference on Microreaction Technology, IMRET 2; Topical Conference Preprints*, pp. 39–44, AIChE, New Orleans, USA, (1998).

[21] Schiemann, G., Cornils, B.; *Chemie und Technologie cyclischer Fluorverbindungen*, pp. 188–193, Ferdinand Enke Verlag, Stuttgart (1969).

[22] Siegmund, G., Schwertfeger, W., Feiring, A., Smart, B., Behr, F., Vogel, H., McKusik, B.; *"Fluorine compounds, organic"*, *Ullmann's Encyclopedia of Industrial Chemistry,* Vol. A11, pp. 349–383,.

[23] Dolby-Glover, L.; Chem. Ind. (London), p. 518 (1986).

[24] Grakauskas, V.; J. Org. Chem. **35,** pp. 723–728, 3 (1970).

[25] Cacace, F., Giacomello, P., Wolf, A. P.; **102,** pp. 3511–3515, 10 (1980).

[26] Conte, L., Gambaretto, G. P., Napoli, M., Fraccaro, C., Legnaro, E.; J. Fluorine Chem. **70,** pp. 175–179 (1995).

[27] Hessel, V., Ehrfeld, W., Haverkamp, V., Löwe, H., Wille, C., Jähnisch, K., Baerns, M., Guber, A.; *"Direct fluorination of toluene using elemental fluorine in gas/liquid microreactors"*, J. Fluorine Chem. submitted for publication (2000).

[28] Nusselt, W.; VDI-Z. **60,** p. 549 (1916).

[29] Nusselt, W.; VDI-Z. **67,** p. 206 (1923).

[30] Danckwerts, P. V., Sharma, M. M.; *"The absorption of carbon dioxide into solutions of alkaline and amines"*, Chem. Eng. (1966) 244–280.

[31] Tontiwachwuthikul, P., Meisen, A., Lim, C. J.; *"CO_2 absorption by NaOH, monoethanolamine and 2-amino-2-methyl-1-propanol solutions in a packed column"*, Chem. Eng. Sci. **47,** pp. 381–390 (1992).

9 Microsystems for Energy Generation

Microreactors for energy generation refer to the generation of hydrogen from a fuel [1, 2], in so-called syngas fuel processors, and the oxidative conversion of hydrogen to yield water, performed in fuel cells [3–5], finally supplying energy. So far, reports about microstructured fuel cells are rare. One reference describes a concept of a miniaturized direct methanol fuel cell. Platelets were made by means of wet chemical etching assisted by thin film technology and plasma polymerization [6]. The ionic conductivities of the 1 µm thin membrane were relatively high. Another reference describes compact designs for hydrogen and direct methanol based fuel cells using newly developed membranes [7]. In the framework of this book no further details about this field will be given.

In contrast, a number of publications deal with the development of miniaturized components for use in small-sized syngas fuel processors. These processors basically consist of a vaporizing unit in order to generate gaseous fuel from hydrocarbon liquids which thereafter are converted to hydrogen and carbon monoxide by partial oxidation and steam reforming reactions or a combination thereof (see Figure 9-1). Autothermal reforming is applied as well. In order to reduce the carbon monoxide content before introducing the hydrogen-rich mixture into the fuel cell, a water-gas shift reaction is usually performed afterwards. As an additional benefit, the hydrogen content is further enriched. Finally, the carbon monoxide content is reduced by a preferential oxidation step.

9.1 Microdevices for Vaporization of Liquid Fuels

Compactness and weight reduction are major issues for proper design of vaporizers for improvement of heat transfer. The key to achieving these targets is to increase the specific

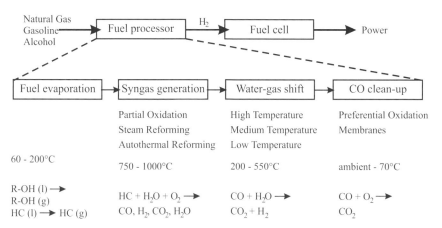

Fig. 9-1. Schematic flow scheme of a system consisting of various components of a syngas fuel processor and a fuel cell.

performance of the evaporating system. The favorable transport characteristics of microdevices render them attractive to be employed in this field. Another driver for miniaturization refers to cost reduction by means of mass production of the individual parts of these microevaporators and automatic assembly to a complete system.

Researchers at the Pacific Northwest National Laboratory (PNNL), Richland (U.S.A.) developed a series of structurally similar microchannel vaporizers based on a sheet-type heat exchanger architecture [8, 9]. Hence, the same operational principle could be utilized and consequently the same sequence of components was employed.

Construction of the Microchannel Vaporizers
The thermal energy for evaporation is provided by catalytic combustion of a dilute hydrogen gas stream from the fuel cell anode effluent using four conventional monoliths (see Figure 9-2). Initiation of combustion is achieved already at room temperature using a palladium catalyst, therefore, requiring no additional preheating unit. After passage through the monolith, the hot combustion gases enter the front side of four underlying sheets comprising microchannel arrays on both sides and finally exit the vaporizer. Thereby, heat is conducted through the sheet base material to the liquid fuel flowing on the rear side. By setting different entrance regions of combustion gases and liquid fuel, a counter-current flow operation can be realized, providing benefits in terms of thermal efficiency.

Due to the structural similarity of the microchannel vaporizers, only for one example, the full-scale gasoline vaporizer, will the dimensions be given. The catalytic monoliths of this device had cross-sections of 1.61 cm^2. The microchannels of the flue gas-side of the sheets were 254 μm wide with an aspect ratio of 18:1. The microchannels of the vaporiz-

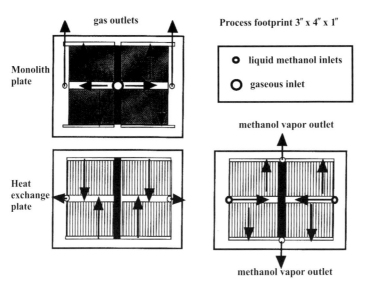

Fig. 9-2. Schematic of a full-scale gasoline vaporizer [8, 9].

ing-side of the sheets were 254 μm wide with an aspect ratio of 10:1. The size of the whole evaporation system amounts to 92 mm x 99 mm x 25.2 mm, the weight being 1.8 kg. The vaporizers were designed for operation with methanol as well as gasoline fuels at either bench- or full-scale size. Using gasoline as fuel, the full-scale microchannel device had to supply sufficient gaseous fuel in order finally to feed a fuel cell of 50 kW power.

Efficiency of Combustion
First functional tests were focused on the catalytic combustion [8]. The feasibility of carrying out the reaction was demonstrated using wet impregnated catalytic powders. Dependent on the hydrogen flow rate, hydrogen conversions typically exceeding 90 % were achieved at pressure drops of 4 to 6 psi and thermal efficiencies of 80 to 90 %. In later tests, Ni monoliths with 60 pores per inch were impregnated using Pd solutions. The hydrogen conversions amounted to 80 to 90 % at pressure drops of up to 20 psi and heat transfer efficiencies of 70 to 90 %, dependent on the methanol flow rate on the vaporizer side.

Heat Transfer Efficiency
The heat transfer efficiencies were further determined as a function of the aspect ratio of the microchannels on the flue gas side (see Figure 9-3). A nearly linear increase in heat transfer efficiency with increasing aspect ratio was recorded. Due to constant overall sheet thickness, the aspect ratio of the liquid fuel microchannels consequently decreased.

The performance tests concerning evaporation will be discussed here only for the full-scale gasoline microchannel vaporizer. An evaporation rate of nearly 300 ml/min using standard gasoline was reached at maximum anode effluent streams of 1400 slpm of a 50 kW

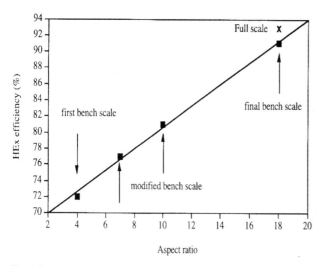

Fig. 9-3. Enhancement of heat transfer efficiency by using flue gas microchannels of various bench- and full-scale vaporizers with a high aspect ratio [8].

fuel cell (6.7 % H_2). The combustion gas stream was heated to 600 °C. The pressure drop on the flue-gas side was about 2 psi, whereas 1 psi was not exceeded on the gasoline side. The testing focused on the determination of the partial-load behavior of the full-scale vaporizer. Even at 25 % gasoline load, corresponding to about 75 ml/min, complete evaporation was achieved. The temperatures measured at the outlet of the vaporizing region were generally higher than 230 °C, exceeding the highest boiling point of the fuel by at least 20 °C.

Similar to the partial load operation of gasoline, the flue gas content was reduced to 5 % of full load. Even when operating both gasoline (25 %) and flue gas (5 %) at minimum partial load, complete evaporation could be achieved. For this operating conditions, a heat flux of about 20 W/cm^2 was measured, which could be increased up to 105 W/cm^2 using superheating conditions.

Based on the size of the microchannel gasoline vaporizer, the authors estimated the size of a complete microchannel fuel processing system to be less than 8 l.

9.2 Microdevices for Conversion of Gaseous Fuels to Syngas by Means of Partial Oxidations

A number of activities concerning realization of miniaturized syngas processors are carried currently being out, e.g., generating hydrogen from methanol or methane fuels by means of reforming or partial oxidation reactions [10–12]. Many of these activities are conducted by joint co-operations of institutes and industry. Therefore, only a small part of the work conducted has been published so far which will be discussed in detail below. A central research target of these activities was the development of microreactors for partial oxidations.

9.2.1 Hydrogen Generation by Partial Oxidations

Partial oxidations are of high interest, in particular for generating hydrogen for proton exchange membrane (PEM) fuel cells. In particular, methane and natural gas are commonly used fuels for this purpose. There are a large number of natural gas feedstocks at various locations due to the natural gas pipelines being widely distributed. Thus, this widespread distribution of gas sources requires reaction units capable of hydrogen production at the point-of-use. Since the existing fixed-bed technology is not suited for miniaturization, microchannel reactors combined with a PEM fuel cell are an interesting approach to render possible stationary energy generation, e.g. at central stations.

A typical characteristic of partial oxidations is their high reaction rate, allowing these processes to be performed in the millisecond range if sufficient mass transport can be achieved. In addition, efficient heat transfer has to be established to prevent hot-spot formation. These features render partial oxidations, in particular, favorable for applications in microreactors. In addition, inherent safety may be guaranteed since the small channels and reaction volumes may behave like integrated flame arrestors and quench explosions. Consequently, most reports about microreactors for energy generation so far referred to partial oxidation reactions.

Due to the high reaction rates, partial oxidations are supposed to have no specific demands on catalyst carrier porosity. Probably, flat noble metal surfaces on microchannel walls are suitable and, ideally, even uncoated construction material may be utilized such as stainless steel containing nickel as internal catalytic material. This minimizes technical expenditure during fabrication of microchannels made of or containing catalyst material.

By parallelization of these channels, the hydrogen output of the microreactor may be varied. Using high space velocities, large throughputs can be achieved within a small, compact volume. Valves may shut down several reaction units or individual plates, thereby allowing a partial load operation.

9.2.2 Partial Oxidation of Methane in a Stacked Stainless Steel Sheet System

At the Pacific Northwest National Laboratory (PNNL), Richland (U.S.A.) a microchannel reactor was developed for the partial oxidation (POX) of methane with air yielding synthesis gas (CO, H_2) in a fast exothermal reaction [10].

$$CH_4 \xrightarrow[\text{Rh/SiO}_2]{O_2} CO + 2\,H_2$$

Rhodium was chosen as catalyst material since this noble metal is known to be especially suited to inhibit undesired coke formation. Therefore, a rhodium nitrate solution was impregnated on porous silica powder, leading to a content of 5 wt.-%. This powder was either packed within the microchannels or placed in the flow distribution zone, referred to as header.

Construction of the Microchannel Reactor
The microreaction system for partial oxidation of methane was based on a planar sheet architecture. Various plates made of stainless steel were stacked on each other, comprising components for combustion, reaction, preheating, quenching, spacer, top and bottom plates. In the process sequence, the combustion of a fuel is used to heat the reaction volume by means of a gas heat exchanger. Remaining heat is used for preheating the reactants in a further heat exchanger. The reaction is thereby initiated and maintained in a microchannel platelet packed with the catalytic powder. In a third heat exchanger operated with air, the product mixture is cooled rapidly.

The stainless steel reaction plate containing 37 parallel channels (see Figure 9-4) was fabricated by means of precision engineering or electro discharge machining (EDM). The platelet size amounts to 70 mm x 38 mm x 4 mm. The 1500 µm deep and 254 µm wide channels are separated by 254 µm wide fins. The catalyst was packed either within the flow distribution zone, referred to as header, or within the first quarter of the microchannels,

Reactants enter   Products travel upward
through the top plate   to the quench section

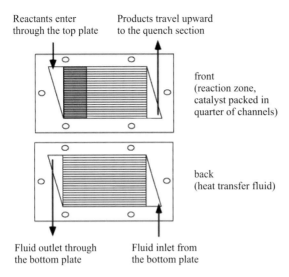

front
(reaction zone,
catalyst packed in
quarter of channels)

back
(heat transfer fluid)

Fluid outlet through   Fluid inlet from
the bottom plate   the bottom plate

Fig. 9-4. Schematic of a stainless steel reaction plate [10].

referred to as channel front section. Thin nickel foils are used for sealing the stacked plates. The temperature was monitored at the header and on the air side of the third heat exchanger. Nickel gaskets were applied as sealing material for the stainless steel plates.

Non-Equilibrium Operation for Catalyst Filled in Header and Channels
Extensive tests were performed on the stacked plate microreactor system using a gas chromatograph for analyzing the reaction products. The experiments had to be limited to a temperature of 700 °C because leakage problems prevented the desired operating temperature of 900 °C from being reached. The inlet feed rate was of the order of some 100 sccm^3, resulting in a residence time of about 50 ms.

The selectivity and the conversion were measured over a temperature range of 300 to 700 °C (channel front section filled with catalyst) and 300 to 600 °C (header filled with catalyst) and compared with the equilibrium values (see Figure 9-5). It turned out that the measured and the equilibrium values differ strongly as to be expected for the non-equilibrium process conditions chosen for these experiments. Practically no coke formation was observed and also the concentrations of combustion products (CO_2 and H_2O) were below the equilibrium values. When the channel front section was filled with the catalyst, the selectivity for carbon monoxide formation increased with temperature and, in contrast to hydrogen selectivity, exceeded the equilibrium values. When the header was filled with the catalyst, nearly constant selectivities of carbon monoxide and hydrogen over the whole temperature range between 300 and 600 °C were found. This was attributed to much higher local temperatures as indicated by the measurements taken at a rather long distance from the reaction zone.

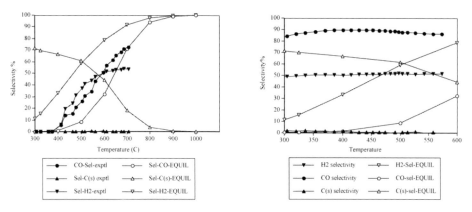

Fig. 9-5 Selectivity for methane partial oxidation: experimental vs. equilibrium [10]. Channel front section filled with catalyst (left), header filled with catalyst (right).

Concluding Remarks

Evidently, the measurements did not cover all aspects of partial oxidation of methane in microreactors, in particular with respect to the local temperature distribution in the reaction zone due to insufficient heat transfer. Nevertheless, it has to be emphasized that these results should be regarded as a first and important step of very promising work in progress on the utilization of microreactors for generation of syngas.

The development of a full operative syngas microreaction unit requires, in addition to a reformer, the realization of pre-reforming, water-gas shift and evaporation components. Future work has to prove higher reactor performance in this field. For instance, maintaining hydrogen output at smaller overall system size is important for both mobile and stationary syngas reformers.

9.2.3 Partial Oxidation of Methane in a Microchannel Reactor

Researchers at the Forschungszentrum Karlsruhe (FZK) (Germany) in cooperation with the Institut für Angewandte Chemie (ACA), Berlin (Germany) realized a microchannel reactor, the central reaction unit being made of rhodium [11].

$$CH_4 \xrightarrow[\text{Rh}]{O_2} CO + 2\,H_2$$

Design Issues and Microfabrication

The experimental setup consisted of a micromixer [13, 14] to permit fast mixing of methane and oxygen. This reactant gas mixture is preheated by guidance through an electrically heated micro heat exchanger (see Figure 9-6) [15]. Since the flow through these two com-

Fig. 9-6. Photograph of a rhodium microchannel reactor and the respective ceramic holder [15].

ponents passes only miniaturized channels, which act similarly to flame arrestors, a safe introduction of a hot reactant gas mixture in the reaction unit is guaranteed.

A central issue concerning the design of a reaction unit for partial oxidations is to ensure efficient heat transfer in order to avoid thermal runaway reactions caused by large releases of reaction enthalpy from total oxidations. The high thermal conductivity of metal microreactors allows one to utilize axial heat dissipation as an efficient heat transport mechanism to avoid hot spot formation. For this purpose, platelets made from rhodium were fabricated by means of mechanical micromachining and thin-wire erosion. Only the wire-eroded platelets were employed during the experimental investigations. The catalyst structure consisted of 23 microstructured platelets, each comprising 28 microchannels, 120 μm wide and 130 μm deep.

Electron beam and diffusion welding were utilized for interconnection of the platelets to a stack. In the latter case, tight interconnection between the platelets could be evidenced by vanishing of the platelet interfaces, being replaced by a coarse grain structure interpenetrating both platelets. The microstructured rhodium catalyst stack is inserted in a ceramic holder to minimize heat losses. By means of a heating wire, ignition temperatures were reached by stepwise increasing the electrical power to 150 W. After reaching a temperature of 1000 °C, the electrical power was reduced stepwise. The reaction was carried out under autothermal conditions. Cooling of the product mixture is achieved by a double pipe heat exchanger.

Conversion and Selectivity of Partial Oxidation of Methane
During experiments, flow rates of 1 to 10 l/min (STP) at pressures ranging from 0.1 to 2.5 MPa were applied. At the beginning of the experiments catalyst activity increased continuously until reaching a constant value after approximately 20 h. In the temperature range of 1090 to 1190 °C maximum conversions of 62 % and 98 % referring to methane and oxygen, respectively, were achieved (see Figure 9-7) The product mixture was mainly composed of hydrogen and carbon monoxide, the corresponding maximum selectivites amounting to 78 % and 92 %, respectively. Similar to the findings of the PNNL researchers [10],

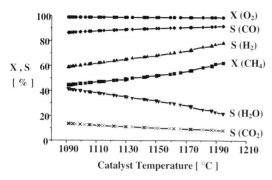

Fig. 9-7. Increase of conversion (X) and selectivity (S) of the partial oxidation of methane as a function of temperature [11].

the work was conducted under non-equilibrium conditions, most likely being due to the short residence times in the range of 1 ms. Reducing the residence time further resulted in product spectra which indicate mass transfer limitations within the microchannels.

Another set of experiments referred to the pressure dependence on product spectra. With increasing pressure up to 1.2 MPa, methane conversion as well as hydrogen and carbon dioxide selectivity decreased. Moreover, soot formation was detected at high pressures, while being absent at low pressure operation.

Concluding Remarks
The experiments of the FZK and ACA researchers elegantly showed a route to exploit microchannel catalytic reactors synthesizing quantitative amounts of hydrogen. The reactor concept is robust, reliable and, especially important for high-temperature reactions, allows safe operation with flammable gas mixtures. As already performed in the framework of the above-mentioned investigations, optimization of microreactor performance by means of CFD simulations can be achieved. The investigations of this second generation of microreactors will bring further valuable insights. Based on such information, the scientific success achieved by using syngas microreactors will soon bring commercial benefits, evidencing their competitiveness with respect to existing small-scale devices realized by conventional techniques.

9.3 Microdevices for Conversion of Gaseous Fuels to Syngas by Means of Steam Reforming

9.3.1 Steam Reforming of Methanol in Microstructured Platelets

Researchers at the Forschungszentrum Karlsruhe (Germany) and of the Universität Erlangen (Germany) investigated the use of nanoparticles as catalysts for microchannel platelets performing the endothermic steam reforming of methanol to generate hydrogen [12].

$$CH_4 \xrightarrow[\substack{Cu/ZnO, \\ Pd}]{H_2O} CO + 2\,H_2$$

A stack consisting of a number of such platelets provides one design concept for a miniaturized methanol reformer. This small, compact system may be utilized, e.g. for automotive applications, since its hydrogen rich gases can be converted on-board into energy by means of a PEM fuel cell. This is of high commercial interest since the first prototypes of cars with conventional tube reformers and fuel cells were developed recently [16]. A disadvantage of these reforming units, apart from their volume and weight, is their slow response time to adapt reactor capacity to varying demand.

Disc-Like Microreactor Encaving Microchannel Platelets
A set-up for methanol reforming in microchannels was developed, consisting of disc-like top and bottom plates encaving several microstructured platelets, being mechanically compressed in a compartment (see Figure 9-8). An evaporator served to generate gaseous methanol from the liquid feed. The reaction zone comprised a stack of the microstructured platelets. The cross-section of the microchannels made of aluminum, separated by 50 µm wide fins, was 100 µm x 100 µm. During experiments four platelets with 80 microchannels and a maximum length of 64 mm were employed. For the operating conditions applied, the residence time in the microchannels amounted to 0.25 s. The platelet stack was electrically heated.

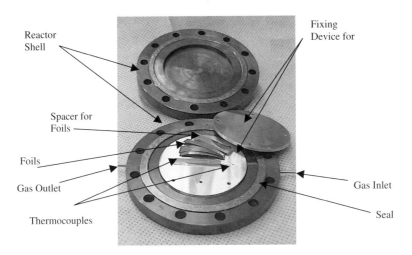

Fig. 9-8. Photograph of a methanol–steam reforming reactor based on microstructured platelets encaved between two disc-like plates [12].

Sintered nanoparticles carrying the catalytically active components were immobilized inside the microchannels. As catalysts Pd and Cu were employed containing ZnO as promoter. In addition, TiO_2 supports were activated by means of wet chemical impregnation.

Porosity of Nanoparticles

Before carrying out the reaction in the set-up, the properties of the nanoparticles were characterized in detail. Temperature-induced dimensional changes of the nanoparticles were monitored by dilatometric measurements, revealing an onset of size reduction at about 400 to 500 °C. BET surface area measurements of the nanoparticles were performed as a function of sintering temperature. With increasing temperature, the BET surface areas decreased from a few tens to a few m^2/g. Consequently, porosity also decreased, ranging from 58 % to 14 %. BET surface areas of the nanoparticle coatings in microchannels were 240 times higher than for uncoated microchannels. By Hg porosimetry measurements only macropores of about 100 nm were identified. The average thickness of the nanoscale layers amounted to 20 μm.

Determination of Activity for Variety of Catalysts

In the course of the experimental investigations on methanol reforming, the activities of several differently treated copper catalysts were determined. It turned out that the highest carbon dioxide yield was achieved with a Cu/ZnO impregnated TiO_2 support (see Figure 9-9). With respect to catalytic activity and stability, palladium-coated platelets exceeded those impregnated with copper. Moreover, the palladium-coated platelets were used to analyze possible limitations by diffusion. Gas velocity was varied, while keeping residence time constant by using platelets of various length and number. Since a comparison of yields obtained during these measurements revealed no distinct difference, no indication of diffusion limitation was found.

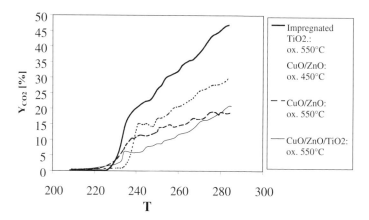

Fig. 9-9. Comparison of activities of various catalysts for steam reforming, as evidenced by their respective carbon dioxide yields [12].

Concluding Remarks

While a number high-temperature microreactors for carrying out partial oxidation of methane have been described, this is one of the rare examples of a miniaturized methanol–steam reforming reactor. The work conducted demonstrated the feasibility of performing methanol reforming in microchannels coated by immobilized catalytic nanoparticles. In addition, details concerning nanoparticle catalyst sintering and a suitable choice of components of optimized catalytic activity are described. This information is interesting on its own since it can be applied to a number of other gas phase reactions in microreactors.

However, in the course of the study it was not clearly outlined which benefits are to be expected in the case of methanol reforming by miniaturization. A clarification of this point is particularly needed since this reforming reaction is a slow, medium-temperature process which should not benefit from transport intensification to the same extent as fast, high-temperature reactions do, such as partial oxidations.

9.4 References

[1] Ridler, D. E., Twigg, M. V.; *"Steam reforming"*, in Twigg, M. V. (Ed.) *Catalyst Handbook*, pp. 225–282, Manson Publishing, London, (1996).

[2] Andrew, S. P. S.; *"The ICI naphtha reforming process"*, in Edeleanu, C. (Ed.) *Materials Technology in Steam Reforming Processes*, pp. 1–10, Pergamon Press, Oxford, (1966).

[3] Ledjeff, K.; *Brennstoffzellen. Entwicklung, Technologie, Anwendungen*, 1st ed; C. F. Müller Verlag, Heidelberg (1995).

[4] Kordesch, K., Simader, G.; *Fuel Cells and Their Applications*, VCH, Weinheim (1996).

[5] Hirschenhofer, J. H., Stauffer, D. B., Engleman, R. R.; *Fuel Cells: A Handbook*, 3rd ed; Gilbert Commonwealth, Inc. for US Department of Energy, Morgantown Energy Technology Center.

[6] Mex, L., Müller, J.; *"Miniaturized direct methanol fuel cell with a plasma polymerized electrolyte membrane"*, in Ehrfeld, W. (Ed.) *Microreaction Technology: 3rd International Conference on Microreaction Technology, Proceedings of IMRET 3*, pp. 402–409, Springer-Verlag, Berlin, (2000).

[7] Hebling, C., Heinzel, A., Golombowski, D., Meyer, T., Müller, M., Zedda, M.; *"Fuel cells for low power applications"*, in Ehrfeld, W. (Ed.) *Microreaction Technology: 3rd International Conference on Microreaction Technology, Proceedings of IMRET 3*, pp. 383–393, Springer-Verlag, Berlin, (2000).

[8] Tonkovich, A. L. Y., Jimenez, D. M., Zilka, J. L., LaMont, M. J., Wang, J., Wegeng, R. S.; *"Microchannel chemical reactors for fuel processing"*, in Ehrfeld, W., Rinard, I. H., Wegeng, R. S. (Eds.) *Process Miniaturization: 2nd International Conference on Microreaction Technology, IMRET 2; Topical Conference Preprints*, pp. 186–195, AIChE, New Orleans, USA, (1998).

[9] Tonkovich, A. L., Fitzgerald, S. P., Zilka, J. L., LaMont, M. J., Wang, Y., VanderWiel, D. P., Wegeng, R.; *"Microchannel chemical reactor for fuel processing applications. – II. Compact fuel vaporization"*, in Ehrfeld, W. (Ed.) *Microreaction Technology: 3rd International Conference on Microreaction Technology, Proceedings of IMRET 3*, pp. 364–371, Springer-Verlag, Berlin, (2000).

[10] Tonkovich, A. L. Y., Zilka, J. L., Powell, M. R., Call, C. J.; *"The catalytic partial oxidation of methane in a microchannel chemical reactor"*, in Ehrfeld, W., Rinard, I. H., Wegeng, R. S. (Eds.) *Process Miniaturization: 2nd International Conference on Microreaction Technology; Topical Conference Preprints*, pp. 45–53, AIChE, New Orleans, USA, (1998).

[11] Mayer, J., Fichtner, M., Wolf, D., Schubert, K.; *"A microstructured reactor for the catalytic partial oxidation of methane to syngas"*, in Ehrfeld, W. (Ed.) *Microreaction Technology: 3rd International Conference on Microreaction Technology, Proceedings of IMRET 3*, pp. 187–196, Springer-Verlag, Berlin, (2000).

[12] Pfeifer, P., Fichtner, M., Schubert, K., Liauw, M. A., Emig, G.; *"Microstructured catalysts for methanol-steam reforming"*, in Ehrfeld, W. (Ed.) *Microreaction Technology: 3rd International Conference on Microreaction Technology, Proceedings of IMRET 3*, pp. 372–382, Springer-Verlag, Berlin, (2000).

[13] Schubert, K., Bier, W., Brandner, J., Fichtner, M., Franz, C., Linder, G.; *"Realization and testing of micro-structure reactors, micro heat exchangers and micromixers for industrial applications in chemical engineering"*, in Ehrfeld, W., Rinard, I. H., Wegeng, R. S. (Eds.) *Process Miniaturization: 2nd International Conference on Microreaction Technology, IMRET 2; Topical Conference Preprints,* pp. 88–95, AIChE, New Orleans, USA, (1998).

[14] Schubert, K.; *"Statischer Mikrovermischer"*, Forschungszentrum Karlsruhe.

[15] Brandner, J., Fichtner, M., Schubert, K.; *"Electrically heated microstructure heat exchangers and reactors"*, in Ehrfeld, W. (Ed.) *Microreaction Technology: 3rd International Conference on Microreaction Technology, Proceedings of IMRET 3,* pp. 607–616, Springer-Verlag, Berlin, (2000).

[16] Ketterer, H.; *"Strom direkt aus Gas"*, Elektronik **23,** p. 152 (1998).

10 Microsystems for Catalyst and Material Screening

Starting from pharmaceutical research with respect to drug identification and optimization [1–3], combinatorial methods for synthesis and screening have become more and more important for other chemical and biological species as well, e.g. regarding homogeneous or heterogeneous catalysts or other types of materials [4–7]. A comprehensive review of this field is far beyond the scope of this book. Therefore, in the following those applications will only be mentioned which explicitly use reactors composed of microstructured parts for a parallel or rapid serial screening. Other numerous approaches which utilize silicon wafer and thin film technology for material screening, e.g. the finding of heterogeneous catalysts or luminescent materials, will not be taken into account here. The reader is referred to reviews providing extensive information on this topic, see e.g. [7]. In addition, the employment of conventional fabrication technologies to build up parallel operated small reactors, although referred to as microreactors [8], will not be discussed in detail.

Specific advantages of using microreactors compared with the existing technologies mentioned above are their unique mass and heat transfer properties as well as the uniform flow distribution, ideally being free of any dead zones. Hence microreactors provide a well-defined setting of operating conditions. Moreover, due to their small volume, fast changes of operating conditions can be performed with minimal time demand to reach equilibrium. The capability of microreactors to test a number of catalysts in separate reaction chambers, without any interference by flow mixing, guarantees high accuracy and reliability of the results, while maintaining a high analysis speed using compact devices.

So far, first microreaction devices for material screening have been built and tested. The feasibility of performing fast analysis was proven [9, 10]. However, a complete characterization in terms of performance, time demand and costs starting from catalyst preparation until analysis is required. This includes the development of practical approaches for fast and reversible filling and removal of catalyst samples into a microreactor. Considering that these activities have been running for no longer than two years, these issues certainly will be addressed in future investigations. Following the huge industrial interest, it is reasonable to predict that the application of microreactors for material screening will gain notably in importance and broaden in near future. For instance, industrially relevant fields of applications such as the screening of polymers and homogeneous catalysts in microreactors are still in their infancy.

10.1 Parallel Screening of Heterogeneous Catalysts in a Microchannel Reactor

Hönicke and Zech at the Universität Chemnitz (Germany) developed in co-operation with the Institut für Mikrotechnik Mainz (Germany) a microchannel reactor module [9, 11]. Their aim was to analyze the activity of heterogeneous catalysts in parallel, coated on a carrier platelet under typical process conditions of gas phase reactions, e.g. demanding

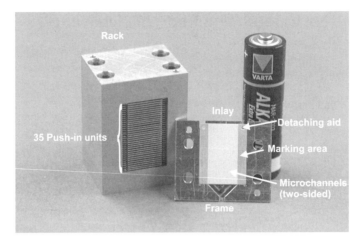

Fig. 10-1. Photograph of a reactor module, comprising 35 stacked frames, which carry catalyst coated wafers, for rapid serial screening of catalysts employed in gas phase reactions [9].

temperature and chemical stability of the module. For analysis of the product mixtures, a newly developed sampling device was used.

Stacked Frame Microchannel Reactor

The module was designed as a stack of 35 stainless steel microstructures, termed frames since they serve for insertion of microstructured platelets carrying the catalytically active material (see Figure 10-1). The platelets are either characterized by an array of parallel microchannels, realized by means of wet chemical etching, or only one channel structure of similar size as the microchannel array, fabricated by means of micromilling. Aluminum was chosen as the platelet material since the surface of this material can be enlarged by anodic oxidation and the corresponding oxide layer can be coated by standard wet chemical methods. The application of thin film techniques or any other type of deposition is also possible.

The frames are high-quality materials with polished surfaces to ensure gas tightness, whereas the catalyst platelets are cheap, disposable units. The separation of reaction unit and sample provides a great degree of freedom concerning these type of sample carrier, material, or deposition technique. The choice of construction materials and sealing concept allows the operation of the microreaction module up to about 450 °C.

Product Sampling

Sampling is achieved by introducing the product mixtures from the frame outlets into a 300 μm wide silica capillary connected to a mass spectrometer (see Figure 10-2). In order to avoid crossover from different outlets, the capillary has to be inserted into the frame outlet, e.g. up to 5 mm. A high accuracy of capillary positioning is needed to ensure a reliable automatic operation of sampling. An x–y positioning system guarantees a preci-

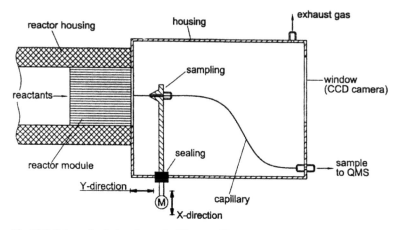

Fig. 10-2. Schematic of a housing embedding a capillary for automated sampling of gas mixtures generated in the microchannel screening reactor [9].

sion of about 5 µm. The sampling system is enclosed by a gas tight housing which contains openings for fluid connection to the frame outlets and mass spectrometer, respectively. The exhaust gases pass this housing and are removed via a third opening.

Signal-to-Noise Ratio and Catalyst Activity of a Library
Before testing the catalytic activity within the module, basic investigations were performed to ensure the feasibility of the chosen concept. A first experiment was undertaken to estimate the response time of the module. This was achieved, for a typical set of operating conditions, by determination of the time delay between setting of a concentration pulse and detection of the response at outlet of the frame. It took about 8 s until a change in sampling signal was evident, and a constant value was achieved after 15 s. The authors estimated that, therefore, cycle times for sampling of at least 60 s are possible. A whole run of all 35 catalyst samples takes not much longer than half an hour.

A second set of experiments referred to the accuracy of sampling. An air flow, simulating "noise", was set in a cross-flow configuration at the product outlet to an argon flow having passed the microreaction module, simulating "signal". Depending on the argon flow rate, sufficiently high signal-to-noise ratios ranging from 70 to about 10,000 were achieved. In addition, it was shown that deep penetration of the capillary into the frame is essential.

After demonstrating conceptual feasibility, a set of 35 different catalyst samples were tested using the microreaction module, comprising various wet chemically deposited metals (Pt, V and Zr) and carrier surface areas (see Figure 10-3). The oxidation of methane at 430 °C and 1.1 bar was used as a reference reaction. In a cycle time of 60 min, distinct differences between catalyst activity of the mixtures were identified. Confidence in these results was based on the good reproducibility of the results (four cycles) and on reasonable structure–property relations, e.g. that high surface area catalysts displayed high activity.

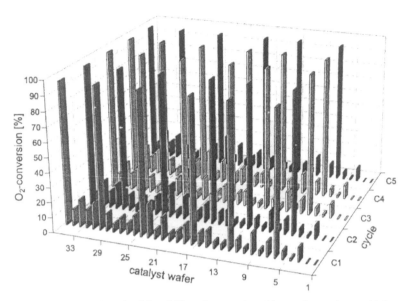

Fig. 10-3 Determination of activity of 35 catalyst samples with regard to methane oxidation carried out in the microchannel screening reactor [9].

$$CH_4 \xrightarrow[\text{Pt,V,Zr}]{O_2} CO_2$$

Concluding Remarks

The straightforward, practical approach discussed above achieved success after a short development time of less than one year. The key was to introduce innovative solutions only in the case of substantial improvement of overall performance and otherwise to rely on existing technology. New developments referred to the microreaction module as well as to the sampling device. A first proof of function was supported by basic conceptual verification. Further measurements should refer to automated catalyst synthesis and sample insertion in order to estimate fully the potential of the concept.

10.2 Parallel Screening of Heterogeneous Catalysts in Conventional Mini-Scale Reactors

Researchers at the Institut für Angewandte Chemie, Berlin (Germany) employed a multitube reactor equipped with 15 quartz tubes of 3.5 mm inner diameter [10]. These tubes were encaved in a stainless steel heating block. Analysis was performed by a quadrupole mass spectrometer. Two 16-port valves at the inlet and outlet of the quartz tubes were used to

Fig. 10-4 Photograph of a capillary moved by an x−y−z positioning system for sampling of gas mixtures leaving a ceramic monolith catalyst carrier [10].

guide feed and product mixture flows. One catalyst test was performed within 5 to 10 min, hence the analysis of one set of 15 catalysts required no longer than a few hours.

Ceramic Monolith Connected to Automated Sampling System
A ceramic monolith reactor, consisting of straight channels of 2 mm inner diameter, was developed to analyze 250 catalyst samples. A stainless steel block served for heating. Sampling was achieved by means of a fused silica capillary moved by an x−y−z positioning system with three stepping motors (see Figure 10-4). Full computer control of the system enabled catalyst screening to be performed automatically. Testing of one catalytic channel required about 1 min. The determination of catalyst activity for all 250 channels hence required only a few hours. The authors mentioned that flow equipartition of the ceramic monolith was reduced with respect to the multitube quartz reactor.

$Mn/Na_2WO_4/SiO_2$ mixtures of varying composition were prepared via the incipient wetness method using an automatic liquid handling system. These samples were tested in the quartz reactor for their catalytic activity. The oxidative coupling of methane was employed as a test reaction. The results obtained were similar to previous findings in a conventional tubular flow reactor.

$$CH_4 \xrightarrow[Mn/Na_2WO_4/SiO_2]{O_2} H_3C\text{-}CH_3$$

10.3 References

[1]   Weber, L.; *"Kombinatorische Chemie – Revolution in der Pharmaforschung?"*, Nachr. Chem. Tech. Lab. **42,** pp. 698–702, 7 (1994).

[2]   Balkenhohl, F., von dem Bussche-Hünnefeld, C., Lansky, A., Zechel, C.; *"Kombinatorische Synthese niedermolekularer organischer Verbindungen"*, Angew. Chem. **108,** pp. 2437–2488 (1996).

[3]   Jung, G., Beck-Sickinger, A. G.; *"Methoden der multiplen Peptidsynthese und ihre Anwendungen"*, Angew. Chem. **104,** pp. 375–500, 4 (1992).

[4]   Cong, P., Doolen, R. D., Fan, Q., Giaquinta, D. M., Guan, S., McFarland, E. W., Poojary, D. M., Self, K., Turner, H. W., Weinberg, W. H.; *"Kombinatorische Parallelsynthese und Hochgeschwindigkeitsrasterung von Heterogenkatalysator-Bibliotheken"*, Angew. Chem. **111,** p. 508, 4 (1999).

[5]   Danielson, E., Golden, J. H., McFarland, E. W., Reaves, C. M., Weinberg, W. H., Wu, X. D.; Nature **389,** pp. 944–948 (1997).

[6]   Gong, P., Doolen, R. D., Fan, Q., Giaquinta, D. M., Guan, S., McFarland, E. W., Poojary, D. M., Self, K., Turner, H. W., Weinberg, W. H.; Angew. Chem. (Int. Ed.) **38,** pp. 484–488 (1999).

[7]   Jandeleit, B., Schaefer, D. J., Powers, T. S., Turner, H. W., Weinberg, W. H.; *"Kombinatorische Materialforschung und Katalyse"*, Angew. Chem. **111,** p. 2649, 17 (1999).

[8]   Senkan, S. M., Ozturk, S.; *"Entdeckung und Optimierung von Heterogenkatalysatoren durch kombinatorische Chemie"*, Angew. Chem. **111,** p. 867, 6 (1999).

[9]   Zech, T., Hönicke, D., Lohf, A., Golbig, K., Richter, T.; *"Simultaneous screening of catalysts in microchannels: methodology and experimental setup"*, in Ehrfeld, W. (Ed.) *Microreaction Technology: 3rd International Conference on Microreaction Technology, Proceedings of IMRET 3,* pp. 260–266, Springer-Verlag, Berlin, (2000).

[10] Rodemerck, U., Ignaszewski, P., Lucas, M., Claus, P., Baerns, M.; *"Parallel synthesis and testing of heterogeneous catalysts"*, in Ehrfeld, W. (Ed.) *Microreaction Technology: 3rd International Conference on Microreaction Technology, Proceedings of IMRET 3,* pp. 287–293, Springer-Verlag, Berlin, (2000).

[11] Ehrfeld, W., Hessel, V., Kiesewalter, S., Löwe, H., Richter, T., Schiewe, J.; *"Implementation of microreaction technology in process engineering and biochemistry"*, in Ehrfeld, W. (Ed.) *Microreaction Technology: 3rd International Conference on Microreaction Technology, Proceedings of IMRET 3,* pp. 14–35, Springer-Verlag, Berlin, (2000).

11 Methodology for Distributed Production

Apart from the use of microreactors as precise measuring and screening tools, their potential for chemical production in terms of process flexibility, capacity, variability, and inherent safety was identified from the very beginning of technology implementation [1–4]. A particular revolutionary feature was that microreactors should allow one, in selected cases, to replace or, more realistically, to supplement existing large-scale plants. One concept, which was termed distributed production or production on-site [5], relied on localized production units capable of producing small amounts of chemicals [1, 4]. A number of arguments for distributed processing were identified, especially regarding process safety and elimination of transport and storage in the case of hazardous chemicals.

When the first conceptual approaches to build up small production units, referred to as miniplants [4, 6, 7] (see also [8]), were made, no specialized microreactors for production were available. Relevant information needed for the design of small production units, e.g. space–time yields, was not accessible at that time. Thus, the visionary theoretical work either had to be based on conventionally fabricated reactor parts [4, 6, 7, 9] or did not outline reactor construction in detail [10]. Despite the lack of a clear numbering-up concept, potential applications of distributed production in microreactors and a strategy for implementation were precisely given.

To date, the importance of this work is highlighted by evidence gained from experiment [11, 12] and estimated calculations [13, 14], that microstructured devices utilizing comparatively large channels are a practicable approach at least for selected production processes. In addition, first microreactor devices with a considerable higher degree of parallelization have been constructed after completion of the conceptual approaches [15]. However, they are still not commercially available in large numbers or at a sufficiently low price.

Due to a current change in technology position and to the fact that microreactors and distributed production are still often mentioned in one breath, an analysis of the methodological approaches for a miniplant design has become a burning issue again.

11.1 The Miniplant Concept

The miniplant concept was been proposed first by Ponton of the University of Edinburgh (U.K.) in 1993 [4, 6, 7, 9]. The potential benefits of distributed processing, as outlined above and reported elsewhere [1, 2], were identified, in particular focusing on an increase in process safety. However, the critical evaluation of the miniplant concept revealed a number of potential drawbacks as well. For instance, the specific production costs of small production units not only depend on reactor miniaturization, but are also determined by control instruments and other peripheral equipment.

A particularly important feature of the work of Ponton is stressing the importance of reducing process complexity of distributed small-scale devices, e.g. for reasons of maintenance and process reliability. Hence, miniplants should not be just a miniaturized reflection of industrial reactors, but resemble more a derivative of this benchmark. This, however, may require the finding of new or at least adapted process parameters when operating microreactors.

As major unit operations contributing to system complexity of large-scale plants separation and recycling were identified, whereas usually less technical expenditure is needed for carrying out a reaction. Hence processes with near exclusive formation of products or relatively simple separation thereof have to be identified. As the type of processing, semi-batch operation with solid feed or product is recommended. The demand for process simplification holds, in addition, for energy supply which has to be guaranteed even at remote locations. For this purpose, the use of electrical heating and cooling is advised. The construction of the equipment should be robust since maintenance may not be possible for longer periods at remote sites. To ensure safety and to minimize pollution, encapsulation of the whole plant is suggested.

11.1.1 Miniplant Concept for HCN Manufacture

As an example of use of the miniplant concept, the Andrussov process yielding hydrogen cyanide (HCN) from methane and ammonia was chosen [9]. This reaction is performed on an industrial scale, the highly toxic product being distributed from a central station by shipping or other transportations means. Miniplants allow distributed production for synthesis of only small quantities to saturate special markets, e.g. supply of HCN for laboratory purposes.

$$CH_4 + NH_3 + 1.5\,O_2 \xrightarrow{\quad Pt \quad} HCN + 3\,H_2O$$

A simple flowsheet for a minaturized process avoiding any complex recycle and separation systems was proposed. Although this processing is not efficient in terms of energy recovery, notable gains in safety and process flexibility still render miniaturization of the Andrussov reaction attractive. This holds even more when considering an alternative chemistry route for HCN synthesis based on the Degussa process (see also [10]).

$$CH_4 + NH_3 \xrightarrow{\quad Pt \quad} HCN + 3\,H_2$$

For this process, a concept of a miniature plant using microwave heating and a fuel cell for energy generation from hydrogen was proposed (see Figure 11-1).

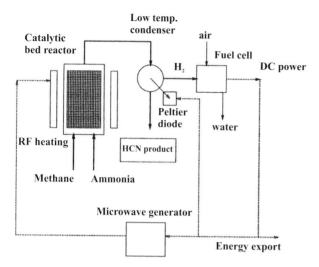

Fig. 11-1. Miniature batch HCN plant using the Degussa route [9].

11.1.2 The Disposable Batch Miniplant

Use of Polymers as Disposable Construction Material
After sketching a general framework for an economic operation of miniplants and exemplarily presenting simple flowsheets thereof, Ponton aimed at developing guidelines for inexpensive plant construction. The idea to use reactors of limited lifetime and made of disposable and recyclable materials, referred to as disposable batch plant [7, 9], was oriented on the highly sophisticated chemical manufacture of living organisms, animals and plants. Ideally, such systems would require no internal cleaning, repair or maintenance.

As a cheap, recyclable construction material for microreactors with sufficient short-term chemical resistance, polymers were explicitly mentioned. A further argument for the use of polymers is that for this material flexible computer-aided rapid prototyping methods are available in order to produce reactor components of complex shapes at moderate cost. The low thermal stability of polymers, however, demands advanced heating concepts when carrying out high-temperature reactions. For example, this may be accomplished by means of microwave heating, avoiding the generation of hot surfaces.

In order to reduce the technical expenditure of processing, the use of pumps for the transport of process fluids should be avoided, and such transport should be simply based on peristalsis or gravity feed, hence, eliminating moving parts. For product analysis sophisticated non-invasive techniques are proposed.

Capacity of a Disposable Plant for HF Production
The disposable batch plant concept was exemplarily verified by a theoretical study concerning distributed HF production. Assuming microwave heating as energy source for the

reaction zone, calculations show that a power of 1 kW would be required for producing 20 kg of aqueous HF at 0.6 kg/h. The reactor volume and the total plant area, thereby derived, amounted to 0.04 m^3 and 0.5 m^2, respectively. The total weight of the plant referred to about 120 kg including chemicals.

$$NaF \xrightarrow{\quad H_2SO_4 \quad} HF$$

The concept of the disposable batch plant was further extended by using robotic systems, moving plant items from station to station, referred to as the table-top pipeless plant. As an example of use for this concept, an ethanol production unit was outlined consisting of a fermenter/stillpot, topped by a packed column section, and a partial reflux condenser.

Concluding Remarks
The miniplant concept has both a visionary part and provides a list of practical guidelines for determination of concrete applications. This two-fold information, although not explicitly developed for microstructured devices, is still valid for future activities in this field. Certainly, the technical transfer of these ideas is not trivial at all and demands further development. Also, the feasibility of a number of assumptions, e.g. concerning simple solutions for pumping fluids, has to be verified. Nevertheless, the work serves as an ideal textbook full of fundamental advises for processing chemistry different from established technology.

11.2 Paradigm Change in Large-Scale Reactor Design Towards Operability and Environmental Aspects Using Miniplants

On the basis of ideas similar to the concept of Ponton [4, 9], a systematic approach, referred to as design methodology, was developed by Rinard of the University College New York (U.S.A.) [10]. Starting from a historical analysis of the main drivers governing plant development, it is suggested that one should change the relative weighting of these drivers. The design of modern plants should be much more oriented according to standards of environment, safety and process control, despite relying solely on productivity. These tasks are ideally met by miniplants being modular assembled from small components. Despite utilizing small existing processing components, miniplant construction using microstructured devices is additionally taken into account.

Historical Analysis of Chemical Plant Development
Rinard starts his analysis by reporting the rapid growth of plants, e.g. in the petrochemical industry, in recent decades and the consequent impact on the environment. This increase in plant size was mainly motivated by economy of scale, i.e. a decrease of specific capital costs. In addition, the efficiency of the plants increased by the use of new technologies, e.g. saving energy as well as advances in instrumentation and process control. Within this his-

tory of technological improvements, disadvantages of using large-scale reactors were figured out, such as potential risks of transport and storage of hazardous chemicals as well as toxic emissions during production. Another drawback of existing equipment, being custom-designed and custom-built, refers to the low flexibility regarding process variation. Hence times for process adaptation or development are long and costly.

These disadvantages should be overcome by miniplants, generally defined similar as in [4, 9], strictly based on a modular design. In order to adapt to varying needs, several modules may be operated in parallel. Typical module capacities range from 100,000 to 1,000,000 lb/yr. Operation of the miniplants should be so reliable and simple that the majority of these plants can be operated by personnel not specially skilled in process technology. Start-up and shutdown have to be performed fast to allow just-in-time production. The entire plant should be transportable including footprint and containment volume.

Safety and Controllability as Major Design Issues
Rinard stresses, in particular, that prior attempts mainly focused on improving process flow sheet rather than rendering priority to safety (see also [16]), controllability (see also [17, 18]), or other operational issues. Existing technology suffers further from the accuracy of simulations which could enable a real conceptual design strategy. In addition, process control is limited, e.g., by the inability to obtain comprehensive information owing to restrictions in the number of control units and for method-related reasons, completing basic design before operability considerations. Environmental aspects are usually addressed thereafter ("end-of-the-pipe" approach). Safety reflections are also performed at a late stage in design.

As a solution for a different type of processing, stressing operability and environmental aspects, a simplicification of chemical processes is suggested, referred to as the KISS ("Keep it simple, stupid") principle. This, in particular, can be achieved by suitable choice of process chemistry and of plant components. As an example of a conceptually simple process, the Degussa route [19] of producing hydrogen cyanide is mentioned (see Figure 11-2). For

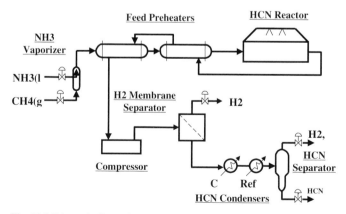

Fig. 11-2 Schematic flow scheme of a miniplant for hydrogen cyanide formation via the Degussa route [10].

this process, a simple miniplant operation is sketched using a microwave oven for energy supply [20], benefiting from simple separation. As suitable process equipment, plant components based on combined operations are introduced, i.e. membrane reactors. For product separation, only sufficiently small components should be employed, hence favoring gas membrane and adsorptive separators over other separation equipment such as distillation.

$$CH_4 + NH_3 \xrightarrow[Pt]{} HCN + 3 H_2$$

Another issue of the miniplant concept relies on standardization, rendering a different strategy for increasing throughput. Whilst current scale-up processes are based on an incremental progression of know-how, this should be achieved in miniplants in one stage. The small standardized modules should, at best, be fabricated using mass production techniques.

Encapsulation in containment vessels allows near zero emission rates to be reached in miniplants, e.g., by purging the reactor with an inert gas sent to a scrubber. This embedding of the miniplant also should dramatically reduce the risk of explosion or environmental contamination in case of an accident. Even if modules of the miniplant are damaged or break, the robust encavement will be mechanically and chemically stable enough to prevent pollution.

Supply-Chain Systems
Considering supply-chain systems, the usage of relatively nonhazardous raw materials is envisaged. Hazardous products should be used only locally and intermediately. If both products and raw materials are hazardous for a given process, it is recommended to include further processes until getting back to non-hazardous raw materials. In this case, an entire sequence of plants is needed. In this context, Rinard outlines a supply-chain system for the manufacture of toluene diisocyanate. Although chlorine is used in the system, it is present in neither the starting raw materials nor the product.

Concluding Remarks
Similarly to the concept of Ponton, the theoretical analysis of Rinard combines visionary and practical aspects. Thus, both references supply a number of useful guidelines for evaluation of processes, especially suited for miniplant operation. The enormous efforts concerning microreactor characterization, currently undertaken world-wide, will lead to a sound and detailed know-how basis of microreactor efficiency and economy. Implementing this information in a systematic comparison of small and large-scale reactor operation, based on a different weighting of economical and ecological drivers, would gain a deep insight into the potential of miniplant processing.

11.3 References

[1] Lerou, J. J., Harold, M. P., Ryley, J., Ashmead, J., O'Brien, T. C., Johnson, M., Perrotto, J., Blaisdell, C. T., Rensi, T. A., Nyquist, J.; *"Microfabricated mini-chemical systems: technical feasibility"*, in Ehrfeld, W. (Ed.) *Microsystem Technology for Chemical and Biological Microreactors,* Vol. 132, pp. 51–69, Verlag Chemie,, Weinheim, (1996).

[2] Ehrfeld, W., Hessel, V., Möbius, H., Richter, T., Russow, K.; *"Potentials and realization of micro reactors"*, in Ehrfeld, W. (Ed.) *Microsystem Technology for Chemical and Biological Microreactors,* Vol. 132, pp. 1–28, Verlag Chemie, Weinheim, (1996).

[3] Wegeng, R. W., Call, C. J., Drost, M. K.; *"Chemical system miniaturization"*, in Proceedings of the "AIChE Spring National Meeting", 25–29 Febr., 1996; pp. 1–13; New Orleans,USA.

[4] Benson, R. S., Ponton, J. W.; *"Process miniaturization – a route to total environmental acceptability?"*, Trans. Ind. Chem. Eng. **71,** p. 160–168, A2 (1993).

[5] Koch, T. A., Krause, K. R., Mehdizadeh, M.; *"Improved safety through distributed manufacturing of hazardous chemicals"*, in Proceedings of the "5th World Congress on Chemical Engineering", 1996; San Diego, CA.

[6] Ponton, J. W.; *"Some thoughts on the batch plant of the future"*, in Proceedings of the "5th World Congress on Chemical Engineering", 1996; San Diego.

[7] Ponton, J. W.; *"The disposable batch plant"*, in Proceedings of the "AiChE National spring meeting,", 1996; New Orleans.

[8] Buschulte, T. K., Heimann, F.; *"Verfahrensentwicklung durch Kombination von Prozeßsimulation und Miniplant-Technik"*, Chem. Ing. Tech. **67,** pp. 718–724, 6 (1995).

[9] Ponton, J. W.; *"Observations on hypothetical miniaturised disposable chemical plant"*, in Ehrfeld, W. (Ed.) *Microreaction Technology, Proceedings of the 1st International Conference on Microreaction Technology; IMRET 1*, pp. 10–19, Springer-Verlag, Berlin, (1997).

[10] Rinard, I. H.; *"Miniplant design methodology"*, in Ehrfeld, W., Rinard, I. H., Wegeng, R. S. (Eds.) *Process Miniaturization: 2nd International Conference on Microreaction Technology; Topical Conference Preprints,* pp. 299–312, AIChE, New Orleans, USA, (1998).

[11] Krummradt, H., Kopp, U., Stoldt, J.; *"Experiences with the use of microreactors in organic synthesis"*, in Ehrfeld, W. (Ed.) *Microreaction Technology: 3rd International Conference on Microreaction Technology, Proceedings of IMRET 3*, pp. 181–186, Springer-Verlag, Berlin, (2000).

[12] Bayer, T., Pysall, D., Wachsen, O.; *"Micro mixing effects in continuous radical polymerization"*, in Ehrfeld, W. (Ed.) *Microreaction Technology: 3rd International Conference on Microreaction Technology, Proceedings of IMRET 3,* pp. 165–170, Springer-Verlag, Berlin, (2000).

[13] Hardt, S., Ehrfeld, W., vanden Bussche, K. M.; *"Strategies for size reduction of microreactors by heat transfer enhancement effects"*, in Proceedings of the "4th International Conference on Microreaction Technology, IMRET 4", pp. 432–440; 5–9 March, 2000; Atlanta, USA.

[14] Veser, G., Friedrich, G., Freygang, M., Zengerle, R.; *"A modular microreactor design for high-temperature catalytic oxidation reactions"*, in Ehrfeld, W. (Ed.) *Microreaction Technology: 3rd International Conference on Microreaction Technology, Proceedings of IMRET 3*, pp. 674–686, Springer-Verlag, Berlin, (2000).

[15] Ehrfeld, W., Hessel, V., Löwe, H.; *"Extending the knowledge base in microfabrication towards themical engineering and fluid dynamic simulation"*, in Proceedings of the "4th International Conference on Microreaction Technology, IMRET 4", pp. 3–22; 5–9 March, 2000; Atlanta, USA.

[16] Kletz, T.; *Plant Design for Safety,* Hemisphere Publishing, New York (1991).

[17] Morari, M.; *"Effect of design on the controllability of continuous plants"*, in Perkins, J. D. (Ed.) *IFAC Workshop: Interactions Between Process Design and Control,* pp. 3–16, Pergamon Press, Oxford, (1992).

[18] Rijnsdorp, J. E., Bekkers, P.; *"Early integration of process and control design, effect of design on the controllability of continuous plants"*, in Perkins, J. D. (Ed.) *IFAC Workshop: Interaction Between Process Design and Process Control,* pp. 17–22, Pergamon Press, Oxford, (1992).

[19] Klenk, H.; *"Cyano compounds, inorganic"*, in Gerhart, W. (Ed.) *Ullmann's Encyclopedia of Industrial Chemistry,* Vol. A8, pp. 159–163, VCH Publishers, New York, (1987).

[20] Wan, J. K. S., Koch, T. A.; *"Application of microwave radiation for the synthesis of hydrogen cyanide"*, in Burka, M. (Ed.) *Proceedings of the Microwave Induced Reactions Workshop,* Vol. A-3–1, EPRI, Palo Alto, CA, (1992).

Index

A

absorber   190
acrylate polymerization   151
advanced silicon etching (ASE)   19
aerosol techniques   175
ammonia absorption   233
analysis systems   4
analytical modules   115, 136
Andrussov process   217, 278
anisotropic wet chemical etching   15
anisotropic wet etching of silicon   17
annular flow pattern   229
anodic oxidation   173, 180
ASE   19
axial diffusion   190
axial heat conduction   104

B

batch systems   4
bifurcation type minimixer   146
breakthrough   102
bubbly flow pattern   229
bulk material   173

C

carbon dioxide absorption   125, 242, 251
catalyst and material screening   271
catalyst carrier   26, 180
catalyst carrying membrane   210
catalyst supply   173
catalytic combustion   133
catalytic dehydration of alcohols   191
catalytic wire microreactor   184
caterpillar mixer   62
ceramic heaters   97
ceramic monolith   275
chemical heat pumps   108 f.
classes of macroscopic mixing equipment   41
closed microcells   169
coatings   33
combustion   259
combustor   32
compact heat exchangers   87, 92
competing reactions   66, 80

condenser   109
continuous flow systems   4
copper oxalate precipitation   165
counter-flow micro heat exchanger   94
corrosion   33
cross-flow heat exchangers   89, 92
cross-flow microfilters   131
cross-type micro heat exchanger   100, 221

D

decrease of diffusion path   80
Degussa route   281
desorber   109
3D GC module   138
die sinking   27
direct fluorination of toluene   247
disc like microreactor   266
disperse gas/liquid systems   229
dispersion   59
disposable batch miniplant   279
distributed production   12, 277
drilling   23
droplet size distribution   63, 80
dry etching of silicon   19
dry etching processes   15
Dushman reaction   162
   – mixing efficiency   163
dynamic response of ceramic heaters   98

E

EDM   26
EDM drilling   29
electrically heated stacked plate devices   97
electrochemical microreactors   166
electro discharge machining (EDM)   26
embossing   24
emulsions   48, 67
entrance flow   106
etching of glass   15, 22
ethylene oxide synthesis   135
evaporator   109
explosion regime   200, 214
externally forced mass transport   83
extraction efficiency   128

extraction of cyclohexanol from water/
cyclohexane   124

F
falling film microreactors   7, 244
falling films   231
fast gas chromatography   136
Fecralloy®   33
Fe(III) ion transfer   117
Fick's law   41
fins   106
flat plate microreactor   237
flow distribution   177
flow equipartition   246
flow multilamination   61
flow patterns   229, 232 f.
fork-like elements   159
fork-like micromixer   58 f.
FOTURAN®   22
fouling prevention   34
frame GC module   137

G
gaseous fuels   260
gas/liquid dispersions   48, 70
gas/liquid microreactors   229
gas phase microreactors   176
gas phase reactions   173
gas purification microsystems   133
gas separation microdevices   134

H
H_2/O_2 reaction   184, 214
 – explosion regime   200
HARM   30 f.
HCN manufacture   278
HCN synthesis   217
heat exchange based on cross-mixing   92
heat exchange performance   91, 96
heat exchanger devices   7
heat exchangers   23, 29
heat transfer correlations   92
heat transfer efficiency   104, 259
height equivalent to a theoretical plate   125
HF production   279
high-energy substreams   52
high-throughput   79
high-throughput mixer   62
honeycomb-like arrangement   73

hydrodynamic focusing   80, 162
hydrogenation of methyl cinnamate   161
hydrogenation of α-methylstyrene   236
hydrogen cyanide synthesis   281
hydrogen generation   260
hydrogen/oxygen reaction   184, 214
 – explosion regime   200, 214
hydrolysis of phenyl chloroacetate   52

I
immobilized nanoparticles   173
impinging jets   7
injection molding   21
in-line IR spectroscopy   136
interconnection techniques   30
interdigital channel configuration   64
interdigital heat exchanger   100
interdigital micromixers   128, 152, 155, 233
isoporous-sieve microfilters   131
isoprene oxidation to citroaconic anhydride
   190
isotropic wet chemical etching   15
 – of metal foils   25

K
ketone reduction using a Grignard reagent
   154

L
laminar flow   43
laminar mixing   42
laser ablation   15
laser micromachining   29
LEGO-system   9
LIGA process   15, 20
liquid phase microreactors   144
liquid phase reactions   143

M
magnetic beads   83
mechanical techniques   23
membrane   29, 135, 214
metal   25, 30
methane conversion   133
micro bubble columns   7, 239
microchannel catalyst structures   177
microchannel reactor   195, 263, 271
micro electro discharge machining   15
microerosive grinding   148

microextractor   29, 115
microfabrication   15
microfilters   130
microfluidizer   52
micro heat exchangers   87
microjet reactor   52
microjets   53
microlamination   30 f.
– of thin metal sheets   30
micromachined membrane   122
micromixer – settler systems   126
micromixers   24, 41
– classes of   43
micromixer/tube reactors   158
micromixing process   151
micromolding   15
micro packed bed reactor   235
microreactors   20
– batch-type   4
– continuous flow systems   4
– definition   1 ff.
– micro total analysis systems (µTAS)   4
– numbering-up   1 ff.
microseparation systems   115
microsystems for energy generation   257
micro total analysis systems (µTAS)   4
milling   23
miniplant concept   277
mixer cascades   55, 58
mixer/catalyst carrier microreactor with
   stacked-plate architecture   205
mixing characterization
– hydrolysis of phenyl chloroacetate   52
mixing efficiency   54, 76
– Dushman reaction   163
mixing nozzle/tube heat exchange
   microreactor   162
mixing quality   63, 67, 80
mixing principles
– macroscopic mixing   41
– miniaturized mixers   43
mixing process
– visualization of   59, 61, 75, 82
mixing tee   49 ff., 232
modular microreactor   197
molecular diffusion   41
monolith reactors   223
multichannel catalyst structure   219
multilamination   236, 239

– in an interdigital channel configuration
   64
– using a stack of platelets   75
– using a stack of platelets with star-
   shaped openings   79
– using a V-type nozzle array   73
multiple flow splitting and recombination   55
– using a ramp-like channel architecture
   62
– using a separation plate   60
multitubular reactor   195

N
nano-scale reactors   5
nitration of benzene   121
non-equilibrium operation   262
nozzle array   73
numbering-up   1 ff., 9 f., 71, 152

O
OAOR process   207
open microcells   169
oxidation of ammonia   209
oxidative dehydrogenation   193

P
pan reactor   195
parallel screening   271, 274
partially overlapping channels   115
partially overlapping microchannels   232
partial oxidation of methane   261, 263
partial oxidation of propene to acrolein   178
partial oxidations   260
PEM   32
PEM fuel cell   260, 266
periodic flow profile   107
periodic operation   188
pilot-scale experiments   155
plate-to-plate electrode configuration   166
potentiodynamic operation   169
precision engineering   15
process development   11, 157
processing, intensification of   11
propane dehydrogenation   192
proton exchange membrane (PEM)   32
punching   24

Q
quench-flow analysis   50

R

radial diffusion   190
reaction systems   4
reactive ion etching processes (RIE)   19
regular flow pattern   182
residence time distribution   188
RIE   19

S

safety issues   11
scaling-up   9
screening   8
segmented flow pattern   230
segmented-flow tubular reactor   164
selective oxidation of 1-butene to maleic
    anhydride   187
selective oxidation of ethylene to ethylene
    oxide   187
selective partial hydrogenation of a cyclic
    triene   180
selective partial hydrogenation of benzene
    186
sieve-like walls   126
silicon   17, 19
sine-wave microchannels   107
slug flow   230
slurry-techniques   175
sol–gel processes   34
sol–gel techniques   173
space–time yields   249, 254
specific interfacial areas   230, 241
split – recombine principle   159
splitting and recombination mechanisms   44 f.
splitting recombination   58
stacked frame microchannel reactor   272
stacked sheet   261
stack of platelets   75, 79
stainless steel   23
star-shaped openings   79
steam reforming of methanol   265

streaming mixers   43
substitution pattern   250
supply-chain systems   282
surface cutting   23
surface-to-volume ratio   7, 33
Suzuki reaction   160
syngas   260
syngas fuel processors   257
synthesis and desymmetrization of thiourea
    159
synthesis of ethylene oxide   203
synthesis of 4-methoxybenzaldehyde   166
synthesis of methyl isocyanate   197
synthesis of microcrystallites   164
synthesis of vitamin precursor   144

T

termed high-aspect-ratio microchannel
    (HARM)   30 f.
thermal blocking structures   105
thermographic measurements   110
thin film coated catalysts   173
throughput   118
transfer efficiency   117, 124
transfer of phenol in octan-1-ol/water   118
T-shaped microreactor   209
turbulent flow   42
turning   23

V

vaporizer   32, 257 f.
visualization of mixing process   59, 61, 75,
    82
vortex formation   177

W

wedge-shaped flow contactor   119
wet chemical etching of glass   22
wire-cut erosion   27
Wittig-Horner-Emmons reaction   161

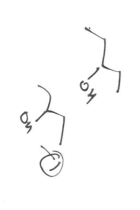

RETURN TO: **CHEMISTRY LIBRARY**

100 Hildebrand Hall • 642-3753

LOAN PERIOD	1	2		3
4		1-MONTH USE	5	6

ALL BOOKS MAY BE RECALLED AFTER 7 DAYS.

Renewable by telephone.

DUE AS STAMPED BELOW.

NON-CIRCULATING
UNTIL: 5/25/01 2pm

MAY 2 0 2005

AUG 1 0 2001

MAY 2 4 2003
MAY 2 4 2003

AUG 1 5 2003

DEC 1 8 2003
DEC 1 9 2006

MAY 2 1

FORM NO. DD 10
3M 3-00

UNIVERSITY OF CALIFORNIA, BERKELEY
Berkeley, California 94720–6000